Philosophy of Biology

Philosophy and Science

Series Editor: Alexander Bird

This new series in the philosophy of science offers fresh treatments of core topics in the theory and methodology of scientific knowledge and introductions to new areas of the discipline. The series also seeks to cover topics in current science that raise significant foundational issues both for scientific theory and for philosophy more generally.

Published

Philosophy of Biology
Brian Garvey

Theories of Scientific Method
Robert Nola and Howard Sankey

Forthcoming titles include

Empiricism
Stathis Psillos

Models and Theories
Roman Frigg

Philosophy of Chemistry
Robin Hendry

Philosophy of Physics
James Ladyman

Psychiatry and Philosophy of Science
Rachel Cooper

Philosophy of Biology

Brian Garvey

ACUMEN

First published in 2007 by Acumen

Acumen Publishing Limited
Stocksfield Hall
Stocksfield
NE43 7TN
www.acumenpublishing.co.uk

ISBN: 978-1-84465-070-5   (hardcover)
ISBN: 978-1-84465-071-2   (paperback)

British Library Cataloguing-in-Publication Data
A catalogue record for this book is available
from the British Library.

Designed and typeset by Kate Williams, Swansea.
Printed and bound by Biddles Ltd., King's Lynn.

Contents

Acknowledgements

I presented some of the material in this book as a visiting speaker at the Department of Philosophy at University College Cork, the Irish Philosophical Club in Ballymascanlon, the Division of History and Philosophy of Science at the University of Leeds, and the Genetics Society at Trinity College Dublin. My thanks go to those institutions for inviting me to speak, and to all who participated in those events. I have benefited greatly from informal conversations with many of my colleagues at Trinity College Dublin and Lancaster University, and with other friends elsewhere. I especially thank Rachel Cooper, Francis Fallon, Richard Gray, Richard Hamilton, Colin Murphy, Jim O'Shea, Gregory Radick, Matthew Ratcliffe, Tomás Ryan, Damien Storey and Dan Watts. I have also taught courses in philosophy of biology at Trinity College Dublin and Lancaster University, which enabled me to test-drive much of the material in this book, and I thank my students for their lively participation and feedback. I thank Alan T. Lloyd for reading drafts of the final two chapters, and for giving me extremely useful commentary. I thank Steven Gerrard of Acumen for his extreme patience with me during the writing and rewriting, and the publisher's readers for extremely helpful and thorough comments. A very special thank you goes to William Lyons, for years of encouragement and mentorship.

Introduction

When I was at primary school, the teacher set my class an exercise of classifying a miscellaneous collection of objects into "alive", "dead" and "never alive". The second group included objects made of wood, paper and leather, and the third included ones made of metal and stone. Even at that early age – we must have been about six at the time – there was no perceived problem in assigning the objects to their various categories. Seeing things as living or non-living seems to be a basic component of our experience. We somehow have a gut feeling that there is an absolute difference between living things and non-living things.

The theory of evolution, however, rides roughshod over this gut feeling. For it leads to the conclusion that the processes going on in living things, and the processes by which they are created, are fundamentally the same as the processes going on in the non-living physical world.

Philosophy of biology can be considered a part of philosophy of science. Very broadly, we might say that philosophy of science deals with questions that arise from science, but are not themselves scientific questions. But this statement needs to be qualified. First, philosophy of science is not the same as history of science or sociology of science, both current growth areas. Nor is it, much, concerned with the ethical questions that arise *because of* science. Medical ethics, for example, is a separate field. Among the questions that philosophers of science ask are questions about the concepts that science uses, questions such as "What is a law of nature?" and "What is a natural kind?" Another question they often ask is "How can we tell real science from pseudo-science?" Secondly, many present-day philosophers see no sharp division between philosophy and science. Philosophy, on this view, is just the more reflective, armchair-speculative end of a spectrum that has hands-on empirical work at its other end. Philosophers who think this – they are usually called "naturalists" – are especially common in philosophy of biology. Philosophers of biology have often worked fruitfully with scientists, and many of the issues in philosophy of biology belong in a zone somewhere between philosophy and science. Indeed, there is no clear boundary between philosophy of biology and theoretical biology. Richard Dawkins considers his book *The Extended Phenotype* (1982) to be a work of philosophy.

Philosophy of science is supposed to be about science in general. But for most of its history it has been much more concerned with the "hard" sciences – physics and chemistry – than any others. The hard sciences were assumed, sometimes tacitly, sometimes openly, to be models of all that a good science should be. Only in the past few decades have distinct subject areas called "philosophy of physics", "philosophy of chemistry" and "philosophy of biology" emerged. (In the case of philosophy of biology, many books and papers have retrospectively become part of the genre.) Partly this is because philosophers have realized that these sciences are different from each other; they do not use all the same concepts, for a start. Even then, though, physics and chemistry seem to go together, with biology the odd one out. Physics and chemistry appear to possess clearly defined entities and strict, exceptionless laws. By contrast, many biological entities have very fuzzy boundaries. Where does a species begin and end, for example? The very idea of a strict, exceptionless law of biology seems misguided. For all that, though, biology is a clear, indisputable example of a science in good working order. It has a well-established core unifying theory – the theory of evolution – and it is precisely because of that theory that there is so much imprecision.

There is conceptual work to be done to understand biology, for it has concepts that seem to belong especially to it. So part of the job of philosophy of biology is to help us to understand these concepts. What is it for something to be innate? What is it for something to have a function? What is a species? I shall address these and other "conceptual" questions in this book. To approach such questions, however, we will first need an understanding of some of the subtle nuances of the theory of evolution, and its place in our overall view of the world.

I shall begin by presenting Darwin's own argument for his theory (Chapter 1). Then I shall show how the theory was modified since Darwin's day to produce a compelling vision of the power of genes (Chapter 2). I shall then look at some themes that have emerged more recently still, such as the question of "units of selection" (Chapter 3), the attack on "panglossianism" (Chapter 4), the emergence of evolutionary developmental biology – "evo-devo" – and developmental systems theory (Chapter 5). Next, I shall address some of the conceptual questions I have mentioned: about innateness, function, biological kinds and species (Chapters 6–9). Philosophers' attempts to answer them have shown how difficult it is to reconcile the often highly counterintuitive consequences of the theory of evolution – not to mention the vast diversity of living examples – with common sense or even simple intelligibility. All this conceptual strangeness is to be found, remember, in a science in perfectly good working order. This will be followed by a chapter on some of the consequences of biology for traditional ideas in philosophy of science (Chapter 10).

The remainder of the book will deal with some of the claims that have been made about what the theory of evolution, and in particular the fact that human beings are products of evolution, implies for various fields of more general concern. There will be discussion of the implications of evolution for episte-mology (Chapter 11), religion (Chapter 12), human nature (Chapter 13) and, finally, ethics (Chapter 14). Clearly, any theory that has, or is claimed to have, implications for such questions as "What can we know?", "Is there a God?", "Who are we?" and "How should we live?" is going to arouse interest far beyond the confines of the academy. The deluge of popular books and magazine arti-cles about evolution and what it tells us is testimony to this. And the passion with which some of these subjects are debated is evidence of much more than intellectual curiosity or love of truth. The battles over evolution versus religion, and over sociobiology and its offspring evolutionary psychology, are battles of hearts as well as minds, and are often fought as if the moral standing, not just the intellectual standing, of the combatants were at stake. As I shall try to show throughout this book, it is important to be clear about what the theory of evolu-tion implies, and, just as important, what it does not imply.

1. The argument in Darwin's *Origin*

What often nowadays goes under the name "Darwin's theory of evolution" is in some ways different from Darwin's actual view. However, present-day evolutionists readily acknowledge that Darwin made the crucial breakthrough. In this chapter I want to show why that is.

Popular and polemical Darwinian literature often contrasts the theory of evolution with religious accounts of the origin of life, that is, with the claim that God is needed to explain the origin of life (e.g. Dawkins 1986). However, it is not at all obvious that the only alternative to Darwinism, as an account of why living things are the way they are, is that God created them that way. More sophisticated defences of Darwinism (e.g. Dennett 1995), sometimes contrast it not only with religious accounts, but with something that they refer to as "essentialism". As this literature presents it, essentialism is, roughly speaking, the view that natural kinds – including biological species – can be given strict definitions, and are fixed and unchanging.[1] It is true that Darwin perceived his assertion that species are not immutable as a major leap in thought: "At last gleams of light have come, and I am almost convinced (quite contrary to the opinion I started with) that species are not (it is like confessing a murder) immutable" (letter to Hooker, 11 January 1844, in Darwin & Seward 1903: 29). But in fact, as he was fully aware, he was not the first person to take this step. Among others, Jean-Baptiste Lamarck (whom I shall discuss briefly here) and Robert Chambers, in his anonymously published *Vestiges of the Natural History of Creation* (1844), took it before him. But that means that Darwin was not the first person to propose a theory of evolution. Why, then, is he regarded as one of the greatest scientists of all time, when Lamarck, Chambers and the rest are not? Darwin's account of evolution, as I shall show in this chapter, is a *successful* account. Unlike earlier evolutionists, Darwin was able to show that the factors that produce evolution actually exist, and are sufficient to explain what he called on them to explain. For the purposes of demonstrating this, I shall very briefly present two earlier accounts of how life got to be the way it is. These accounts are not religious or "essentialist", but attempt to explain things in terms of purely physical processes.[2]

1.1 Earlier attempts

First, a brief word on *what* Darwin called on those factors to explain. Ever since Aristotle, biologists had been aware of two salient facts about living things in general: teleology and taxonomy. *Teleology* (from the Greek *telos*, meaning "purpose") refers to the fact that living things are organized into parts that look as though they were designed for a task. Our eyes are very good at seeing, and our hearts are very good at pumping blood around our bodies. Moreover, both our eyes and our hearts help us to stay alive, that is, they serve the living organism as a whole. It is not surprising that facts such as these have led people to believe that there just *has to be* a God: that they shout out "intelligent design". One might alternatively think that organisms being that way is just a part of the way things are: it does not require any special explanation.[3] Even before Darwin, however, there were people who would not accept either of these things.

Taxonomy (from the Greek *taxis*, meaning "order" or "arrangement") refers to the fact that organisms fall into what appear to be reasonably well-defined classes or categories. Moreover, the categories seem to be *nested*: sheep and wolves are both mammals; mammals and fish are both animals; and so on. Again, one might see this as evidence of God's design, or think that it is just the way things are. Again, not everyone was willing to accept either of these.

The germ of Darwin's idea can be found in ancient times, particularly in the writings of Lucretius in the first century BCE. Lucretius was a member of the Epicurean school of philosophers, which means that his views were in some ways similar to modern-day materialists. Like them, he seems to have wanted to explain how animals are apparently well designed without appealing to an intelligent, deliberate designer. What he proposed was that nature initially produced all kinds of creatures: with wings, without wings, without mouths or any means of feeding themselves, and so on. Some of these simply died because they were ill equipped, while the better-equipped ones survived:

> And in the ages after monsters died,
> Perforce there perished many a stock, unable
> By propagation to forge a progeny.
> For whatsoever creatures thou beholdest
> Breathing the breath of life, the same have been

Even from their earliest age preserved alive
By cunning, or by valour, or at least
By speed of foot or wing.   (*On the Nature of Things*: V)

This might seem to go some way towards an explanation of how creatures come to be well equipped for life. If we want an account of how we get organisms that are apparently designed, without having to invoke an intelligent designer, then Lucretius's story has something going for it. The initial process that produces organisms does not have to be in any way geared to producing ones that are well designed. We should not be surprised that the ones that survived bear the appearance of good design.

It is, however, unsatisfactory on at least one important count. It requires us to believe that, at some time in the past, nature just spontaneously produced large numbers of creatures of various different kinds, completely at random. But we do not see this happening anywhere in the world now. We have never seen nature produce an organism, even a badly designed one, out of non-living matter. Did it do so once, and has it now stopped? Perhaps, but we have absolutely no reason to believe that this ever actually happened. By contrast, as we shall see, the factors that Darwin uses to explain things are processes that we can see happening in the world. Lucretius's theory is thus speculative in a way that Darwin's is not.

One of Darwin's most famous immediate predecessors was Jean-Baptiste Lamarck. Nowadays, Darwin's theory is often contrasted with Lamarck's, but why is Darwin's theory successful and Lamarck's not? Lamarck's name is preserved in the term "Lamarckian inheritance", which is generally used to mean "inheritance of acquired traits". This usage suggests that the difference between Darwin's and Lamarck's theories is that Darwin did not believe in inheritance of acquired traits. This, however, can be shown to be false by his assertion in *On the Origin of Species*: "I think there can be little doubt that use in our domestic animals strengthens and enlarges certain parts, and disuse diminishes them; and that such modifications are inherited" ([1859] 1968: 175).

The real difference between Darwin and Lamarck on this score is the *use* to which each puts the principle of inheritance of acquired traits in his theory. For Lamarck, it is the primary factor that explains how organisms come to be well designed for living in their environments. To use the example for which Lamarck has become notorious, giraffes' necks are said by him to be so long because, in the past, giraffes stretched to reach leaves higher up the trees, and their offspring then inherited their stretched necks. This idea can be given some plausibility if one thinks about the thick skin human beings are born with on the soles of our feet. As we all know, skin gets thicker from continuous friction; guitarists often have thickened skin on parts of their fingers, and violinists on the part of their neck where they rest the violin. We might think, then, that the skin on our ancestors' soles got thicker from years of walking, and that we inherited thick-skinned soles as a result.

But a problem with Lamarck's account is that creatures acquire *many* modifications as a result of the vicissitudes of their lives. Creatures sometimes lose an eye, but their offspring are not born one-eyed as a result. For Lamarck's account to work, there would have to be a systematic bias in inheritance in favour of beneficial traits. It would have to be the case that only they, and not the harmful ones, were inherited; otherwise the harmful ones could accumulate just as much as the beneficial ones. Although Darwin thought that some acquired traits were inherited, he saw no evidence that there was such a bias. His own theory does not require any such bias. Moreover, it does not require inheritance of acquired traits at all, as he himself was aware. Most present-day biologists reject inheritance of acquired traits outright.

Note that Darwin's criticism is not that Lamarck failed to provide a workable mechanism for how acquired traits might be inherited. As we shall see, Darwin was fully aware that he himself knew of no mechanism for inheritance, or for the production of variations. It was sufficient for his purposes that his theory appealed to factors that *could be shown to really exist*, even though he lacked an explanation for them. This is why his theory is successful where Lamarck's is not.[4]

There are further differences between Lamarck and Darwin. Lamarck believed in an inherent tendency to progress and a role for "slow willing of animals". That Darwin criticized Lamarck for this view does at least have textual support, albeit in private letters rather than published works. Immediately after Darwin's statement about species not being immutable, he writes: "Heaven forfend me from Lamarck nonsense of a 'tendency to progression,' 'adaptations from the slow willing of animals,' etc.! But the conclusions I am led to are not widely different from his; though the means of change are wholly so" (letter to Hooker, 11 January 1844, in Darwin & Seward 1903: 29). Once again, Darwin saw no evidence that there was any such inherent tendency to progress, or that animals could, by "slow willing" influence the future direction of evolution. Because of these features, as well as the need for a bias in inheritance, Lamarck's theory can be considered a disimprovement on Lucretius's theory. Lucretius's theory did not require that there be any inbuilt bias towards good design or progress. He could explain how completely unbiased, undirected processes could produce good design.

However, both Lucretius's and Lamarck's theories fall down in virtue of having to appeal to factors for which we have no evidence. Darwin's theory succeeds by appealing to factors that we can see really exist.

1.2 Variation and inheritance

Darwin uses the first chapter of *Origin* to write about the variation of plants and animals under domestication. He describes the activities of cattle-breeders, dog-breeders, pigeon-fanciers[5] and so on, noting how they are able to produce new varieties simply by allowing some individuals to breed and not others. If a dog-breeder wants to produce dogs that have longer legs, she proceeds by taking the longest-legged dogs she has and allowing them to breed, and then taking the longest-legged offspring of those and allowing them to breed and so on. This tried and trusted method, Darwin points out, has produced a wide variety of different kinds of dog, pigeon and so on. In fact, dogs come in so many different shapes and sizes that even expert breeders at the time thought that they must be descended from a number of different wild species, albeit possibly a much smaller number than the great number of different breeds that we have today.

Why, we might ask, does Darwin bother talking about artificially-bred animals? Surely, we might say, (i) Darwin is just telling us something we already know, and moreover (ii) breeders may have produced new varieties of dogs or pigeons, but these are merely *new varieties of old species*. How is this supposed to tell us anything about how species themselves originate? But the title of Darwin's book is *On the Origin of Species*, not *On the Origin of Varieties*. Let us examine these two objections more closely.

(i) There is a world of difference between already knowing something and seeing what it implies. As we shall see, most of the evidence Darwin offers in support of his theory consists of facts that are similarly obvious, or easily derivable from what is obvious. Darwin was careful to amass evidence to verify these facts. For example, he spent some time consulting with professional animal- and plant-breeders. This is good scientific practice, for what is obvious is not always true. The facts that form the basis of his justification for his theory are none of them very novel or surprising. By contrast, think of the following example. According to Einstein's general theory of relativity, gravity acts on light, so that a ray of light bends when it passes close to an object with a large mass, such as the sun. According to Newton's theory, light always travels in straight lines. Thus, the two theories make different predictions. To test the predictions, scientists had to wait for a solar eclipse, to see if the apparent position of stars near the sun in the sky was as Einstein's theory predicted, or as Newton's theory predicted. There was a solar eclipse in 1919, and a team led by the astronomer Arthur Eddington went to Africa to make the relevant observations. Einstein's theory passed the test. Before Einstein developed his theory, nobody expected to find this effect. In that sense, Eddington's discovery is surprising in a way that the empirical facts that Darwin presents about artificial selection are not, and were not even at the time. But empirical facts that are not very surprising are still empirical facts for all that.

(ii) Darwin argues that whether something is a new species or merely a new variety is, in part at least, in the eye of the beholder:

> How many of those birds and insects in North America and Europe, which differ very slightly from each other, have been ranked by one eminent naturalist as undoubted species, and by others as varieties, or, as they are often called, geographical races! Many years ago, when comparing, and seeing others compare, the birds from the separate islands of the Galapagos Archipelago, both one with another, and with those from the American mainland, I was much struck how entirely vague and arbitrary is the distinction between species and varieties. ([1859] 1968: 104)

Ultimately, what he wants to argue is that the transition from one species to another is always by tiny steps, gradually accumulated. The fact that it is not clear when we should say we have a new species, is to be used as a piece of evidence for this.

Moreover, Darwin later sets out to show that natural selection is much more powerful than artificial selection. Later in this chapter I shall get on to how he argues for this, but for the moment I shall stick to what he intends to show by talking about artificial selection.

The production of new varieties by artificial selection is made possible by two factors:

- A great many traits possessed by parents are passed on with a high degree of reliability to offspring.
- This, however, is only *a high degree* of reliability, not absolute reliability. Two parents *may* produce an offspring that (e.g.) has longer legs than either.

Note that *both* of these factors are required for artificial selection to work. If the first were not the case, there would be no assurance that choosing the long-legged individuals for breeding would be any more likely to produce the desired outcome than choosing the short-legged ones. Breeders want the animals and plants they produce to breed true. But if the second were not the case, the production of a new variety with longer legs than ever before would not be possible. The term "natural selection" is often used to include the process whereby the *average* (say) leg length in a population goes up from generation to generation, even if there are no individuals in the new generation with legs longer than the longest-legged individuals in the previous generation. But Darwin explicitly speaks of the production of *novel forms* by artificial selection, and he offers evidence from (among other things) the world of pigeon-fancying to support this. If one was not keeping a careful eye out, one might think that the traits of an offspring are simply a new combination of traits already possessed by its parents. However, Darwin appeals to the testimony of animal- and plant-breeders – whose job requires keeping a careful eye out for such things – to show that this is not always the case.

This raises the question: what explains these two factors? That is, (a) why do organisms by and large breed true and (b) why do novelties sometimes arise? Darwin has no answer to these questions.

As regards (a), the fact that we do not know how to explain it – or, at any rate, they did not know in Darwin's day – should not deter us from affirming that true-breeding is a fact of life. We expect children to resemble their parents. Nowadays, it is usual to explain true-breeding by appeal to the high "copying fidelity"' of genes. Not everyone accepts this, however, and I shall take this issue up in Chapter 5.

As regards (b), as far as Darwin was concerned, the arising of new forms was a free gift of nature. A loose way of describing this situation is to say that mutations are "random". This word, however, carries the connotation that they are events without any prior cause. This is misleading, as there is no reason to believe that they have no prior cause, and Darwin's theory would not be in the slightest bit affected if we discovered that they were fully determined. Darwin comments: "I have hitherto sometimes spoken as if the variations … had been due to chance. This, of course, is a wholly incorrect expression, but it serves to acknowledge plainly our ignorance of the cause of each particular variation" (*ibid.*: 173).

What "random" should be taken to mean in this context is that mutations are *neutral with regard to selection*. To illustrate this point, imagine an animal-breeder who wants to produce cattle with a higher milk yield than previously. The breeder will select cows with the highest milk yield to breed, and in doing so is likely to produce a generation with a higher average milk yield than the last. This need not mean that any of the cows in the new generation have an *unprecedentedly* high milk yield. However, a new individual with an unprecedentedly high milk yield is a possibility. But it is also a possibility that a new individual might arise with an unprecedentedly *low* milk yield, not to mention a new individual with longer horns, or a different coloured coat, or any number of new variations in which the breeder is not interested. This is the sense in which mutations are random; they do not oblige the breeder by being in the direction that she desires.

To sum up, then, Darwin pointed to the fact that creatures tend to inherit traits from their parents. Obvious though this might seem, Darwin went to the trouble of gathering evidence to prove it. Secondly, he pointed to the fact that creatures vary. That is, they can have traits that they did not inherit from their parents.

1.3 The struggle for existence

After reaching these conclusions based on the effects of selection under domestication, Darwin moves on to some observations about the rates at which various organisms reproduce. He drew on Thomas Malthus's famous (or notorious) *Essay on the Principle of Population* (1798). In this work, Malthus had argued that there was a natural tendency for the human population to increase at a faster rate than our ability to produce food increased. Specifically, Malthus thought that population tended to increase *geometrically* – that is, in the ratio 1:2:4:8 and so on – while food production tended to increase only *arithmetically* – that is, in the ratio 1:2:3:4 and so on. If the former tendency went unchecked, then, the level of hunger and concomitant misery would increase proportionally.[6] The best solution would be for people to practise restraint, making sure that they restricted the number of offspring they produced. Malthus thought this was very unlikely to happen, however. Since he thought that giving aid to the hungry would only make the problem worse in the long term, he thought that the only feasible solution was to let nature take its course: let the poorest members of society starve, heartless though it might seem.[7] Darwin says of his own view as regards population increase:

> It is the doctrine of Malthus applied with manifold force to the whole animal and vegetable kingdoms; for in this case there can be no artificial increase of food, and no prudential restraint from marriage. Although some species may be now increasing, more or less rapidly, in numbers, all cannot do so, for the world would not hold them.
>
> ([1859] 1968: 117)

Whatever the merits of Malthus's claims, Darwin gives a simple mathematical argument about the rates at which organisms reproduce. Imagine that we start off with just two individuals of a given species. Imagine that they produce offspring, and that all those offspring produce offspring, and so on. Provided that the average number of offspring produced by any pair is more than two, the population of that species will, as a matter of simple mathematics, increase. If this process continues, sooner or later the resources on which the species depends to live – food, sunlight, nutrients in the soil, space to live in, and so on – will start to come under pressure. There is simply not an infinite amount of food, so the time when there is just not enough to go round has to come sooner or later. Even sunlight is a limited resource, not because there is not enough of it being produced by the sun, but because places in the sun are limited – think of tightly packed trees in a forest putting each other in the shade. Thus it will not be possible for the population to keep increasing indefinitely.

What about the initial condition on which this hypothetical scenario is based: that the average number of offspring produced by any pair is more than two? According to Darwin, we have very good reason to believe that this condition is in fact met by every sexually reproducing species in existence. To make this vivid, he asks us to consider the elephant:

> The elephant is reckoned to be the slowest breeder of all known animals, and I have taken some pains to estimate its rate of natural increase: it will be under the mark to assume that it breeds when thirty years old, and goes on breeding till ninety years old, bringing forth three pairs of young in this interval; if this were so, at the end of the fifth century, there would be alive fifteen million elephants, descended from the first pair.
> (*Ibid.*)

Even though the simple mathematical argument seems perfectly sufficient for his purposes, Darwin also supports his claim empirically:

> [I]f the statements of the rate of increase of slow-breeding cattle and horses in South-America, and latterly in Australia, had not been well authenticated, they would have been quite incredible. So it is with plants: cases could be given of introduced plants which have become common throughout whole islands in a period of less than ten years.
> (*Ibid.*: 118)[8]

So if they are left to breed unchecked, all creatures – even the slow-reproducing elephants – will eventually find themselves faced with a shortage of some resource or other. Once a resource that a species needs to live starts to come under pressure, what Darwin calls "the struggle for existence" ensues. This phrase perhaps conjures up images of creatures fighting each other: "nature red in tooth and claw", as Tennyson put it. There are indeed many instances of creatures fighting each other for valuable resources, for example, birds fighting for territory. In fact, Darwin uses the term "struggle for existence" in a much broader sense than this. The animal that runs faster may be the one that gets to eat; the tree that grows taller may be the one that gets the sunlight.

As the latter example ought to make clear, the term "struggle for existence" should not be taken as implying any intention on an organism's part. Nor need it be confined to struggles between members of the same species, although such struggles will often be the most intense, as you are more likely to need exactly the same resources as a member of your own species than to need exactly the same resources as some other type of creature. If a cheetah is chasing an antelope, they may be said to be struggling for the resource that is the antelope's own flesh, which, after all, the antelope needs to stay alive. But since the antelope that runs slowest will be the one most likely – *ceteris paribus*[9] – to be caught by the cheetah, by running the antelopes may be said to be struggling against each other too. Darwin even says that a plant on its own at the edge of a desert may be said to be struggling for existence, even though the only things that threaten its resources are purely inanimate physical conditions around it: "Two canine animals in a time of dearth, may be truly said to struggle with each other which shall get food and live. But a plant on the edge of a desert is said to struggle against the drought, though more properly it should be said to be dependent on the moisture" (*Ibid.*: 116). Once again Darwin appeals to empirical facts, but facts that, once they are pointed out, have the air of being rather obvious. And once again, in support of these facts Darwin adduces the wisdom of experts, as well as evidence he has gathered himself, and makes fairly straightforward extrapolations from the evidence gathered in these ways.

So now Darwin has established that: (i) organisms inherit traits from their parents; but (ii) organisms vary, (iii) populations of organisms have a natural tendency to increase and (iv) as a result of this there will come times when there is not enough of some resource for all the creatures in a population.

1.4 Natural selection

The next step is to imagine a population of some species that is in the condition that at least one of the resources it needs in order to survive is under pressure. (In modern parlance, this type of situation is called "selection pressure".) At any given time the individuals in the population will not all be the same; there will be some variation. Since there is not enough of something – say, food – to go around, at least some individuals will die before their time. But there is likely to be some difference in how well or badly different individuals fare. Some will suffer less than others.

But what makes the difference? Some of this may be simply luck; a creature may die just because it happened to be in the wrong place at the wrong time. But some of the differences between individual organisms may also make a difference. (Note that it is not necessary for Darwin's argument that *all* differences make this difference.) Among the things that may vary between individuals of the same species, and have been the objects of selective breeding by human beings, are running speed, acuity of eyesight or hearing and production of milk. Clearly, a difference in any of these can make a difference to how a creature fares in the struggle for existence. Faster running speed, or more acute eyesight or hearing may make a creature more likely to catch prey, or escape becoming prey; higher milk yield may mean that more offspring will reach the age at which they can reproduce, and/or will be stronger when they do so. Clearly, the more offspring you have, the greater the chance that some of them will do well in the struggle. If selection pressure is sufficiently intense, even very small differences could be crucial. But if there is any selection pressure at all, then some organisms will do better than others, and this is unlikely to be *entirely* due to luck.

Darwin was strongly committed to the view that only very small variations were involved in this process. As the flipside to this, he thought that the struggle for existence was more or less constant. His overall argument does not actually require the latter to be true. But why should the former be so? The answer is because at any given time an organism is likely to be finely adapted to a particular lifestyle in a particular environment (unless, of course, that environment has just undergone some major relevant change). Moreover, an organism is internally an extremely finely tuned mechanism. Recall also that, as far as Darwin is concerned, the production of variations is random in the sense defined above, that is, it is not intrinsically directed towards the production of desired or favourable outcomes. The finely tuned adaptation of organisms to their environment means that the bigger a random deviation from the norm an individual exhibits, the more likely that deviation is to be harmful rather than helpful to that individual. Think of a complex, finely tuned mechanism such as a television or a car engine. Imagine that a highly incompetent mechanic tries to improve it by randomly messing about with this or that part. Because it is a complex and finely tuned mechanism, the incompetent mechanic is more likely to do harm than good. However, if he or she makes only a very small adjustment, there is a better chance that it might, by a fluke, actually produce an improvement; and at least it is less likely to do harm. The analogue for the fine-tunedness of the mechanism is the fine-tunedness of the organism to its lifestyle in its environment. Mess around with it too much, and you are likely to only do harm; mess around with a little bit, and you are less likely to do harm, and you might even do some good. The same argument applies if we think of the *internal* complexity and fine-tunedness of an organism. A very small adjustment might produce an improvement; a larger one is much more likely to produce a disimprovement, and indeed to produce an organism that is simply not viable. There are many more ways for a complex system to go wrong than for it to go right.

So there will inevitably be situations where not every individual in a population can reproduce to its full capacity, and where some traits possessed by individuals can determine whether an individual reproduces more or less than its peers or, in cases of more intense selection pressure, whether it reproduces at all. The process whereby some individuals are favoured in the reproductive stakes by virtue of traits that they possess that differentiate them from some of their peers is called *natural selection*.

Darwin points out the analogies between natural and artificial selection. In both cases, we have individuals whose traits vary at random. In both cases, some of these variations are advantageous to the individual, in the sense of enhancing its chances of having offspring. In artificial selection this is because the traits are desired by the breeder; in natural selection it is because they enable the organism to be more successful in the struggle for existence.

Now combine this with what we saw in the previous sections: at least some variable traits are *heritable*, that is, passed down to offspring. This means that, just as in artificial selection, so also in natural selection; it is not purely a matter of chance which traits will be more common in the next generation. In artificial selection, if a breeder desires dogs that run faster, and knows what she is doing, then we would expect the average running speed of dogs in the next generation to be higher. In natural selection, if being able to run faster is an advantage to an organism – say, because it helps it to escape from predators – then we would also expect the average running speed in the next generation to be higher. This is provided, of course, that the relevant trait is variable in the population, and that the variations are heritable: that is, that fast-running dogs tend to produce fast-running offspring. But Darwin has already argued from the evidence of plant- and animal-breeders that that is the case for many traits in many organisms.

However, all this shows is that traits that are unfavourable will be weeded out from generation to generation. Slower-running antelope are less likely to survive and produce offspring, so there will be fewer of them in the next generation. This still does not show us how evolution produces more than minor modifications. We are supposed to believe that we are descended from creatures that, for example, had no eyes or lungs. How can eyes, lungs and all the rest, have been produced by this process? To answer this, Darwin argues that natural selection is far more powerful than artificial selection.

The dog-breeders, pigeon-fanciers and so on with whom Darwin consulted were masters of their art. As Darwin says:

> Not one man in a thousand has accuracy of eye and judgement sufficient to become an eminent breeder. If gifted with these qualities, and he studies his subject for years, and devotes his lifetime to it with indomitable perseverance, he will succeed, and may make great improvements; if he wants any of these qualities, he will assuredly fail. Few would readily believe in the natural capacity and years of practice requisite to become even a skilful pigeon-fancier. ([1859] 1968: 91)

But even these master breeders had not produced a creature with eyes from one without eyes, or a creature with lungs from one without lungs. How, then, is it that natural selection can do these things?

One answer that Darwin gave is that even the master breeder with the most skilled eye in the world can only select for things that that eye *can actually see*. Natural selection is not so confined. *Any* trait, however hidden from view, can be acted on by natural selection, as long as it makes a difference to the survival and reproductive prospects of an organism. Moreover, as Darwin continually emphasized, natural selection has plenty of time at its disposal. Even the most artificial varieties of pigeons were produced by selective breeding from the wild varieties well within the period of recorded history. But, as people were beginning to appreciate in the nineteenth century, the age of the earth, and the period during which living things have existed, is far longer than this. Darwin notes that hardened professional animal-breeders believed that there were strict limits on what selective breeding could produce. But this, he argues, betrays a failure of imagination on their part. Evolution, on Darwin's view, was always an extremely slow process. His last book was on the production of the soil of England by the action of earthworms. Stephen Jay Gould (2002: 148–9) has argued that this should be seen as of a piece with Darwin's other projects; he was keen to impress on his readers the capacity of very many small changes to cumulatively produce very big results. In the chapter on natural selection in *Origin*, he waxes lyrical on its superior power compared to human breeders:

> As man can produce and certainly has produced a great result by his methodical and unconscious means of selection, what may not nature effect? Man can act only on external and visible characters: nature cares nothing for appearances, except in so far as they may be useful to any being. She can act on every internal organ, on every shade of constitutional difference, on the whole machinery of life. ... Under nature, the slightest difference of structure or constitution may well turn the nicely-balanced scale in the struggle for life, and so be preserved. How fleeting are the wishes and efforts of man! how short his time! and consequently how poor will his products be, compared with those accumulated by nature during whole geological periods. Can we wonder, then, that nature's productions should be far 'truer' in character than man's productions; that they should be infinitely better adapted to the most complex conditions of life, and should plainly bear the stamp of far higher workmanship? ([1859] 1968: 132–3)

The accumulation of very small changes is crucial to Darwin's theory. But what evidence do we have that this succession of small changes has taken place? We do not have direct evidence, for we have not been observing any species for long enough to catch any evolution actually happening,[10] and the fossil record that we have so far unearthed is not nearly complete enough. Darwin's reasons for believing in cumulative selection, then, involve extrapolation from observed facts and appeal to its being the best explanation.

From a set of conditions – (i) inheritance, (ii) variation, (iii) a tendency for populations to increase if unchecked and (iv) limitations on resources – Darwin *extrapolated* that there will be natural selection, which in turn will produce improvements in the suitability of organisms to live in their environments. Moreover, he showed that these conditions are actually met in the real world. He was able to *explain* the two salient general facts of biology: teleology and taxonomy. His theory explains *teleology* because it gives an account of how traits that fit an organism to live in its environment can arise by means of processes that are not in themselves directed towards that end. In this respect, Darwin's theory is superior to Lamarck's in the same way that I argued that Lucretius's theory is superior to Lamarck's. However, Darwin's theory is also superior to Lucretius's, in that Lucretius produced no evidence that the factors he called on to do the explaining actually existed. As for *taxonomy*, Darwin's theory explains this by being a theory of *gradual modification*. Creatures acquire their distinctive features by natural selection, which modifies them and produces novel forms. This means that creatures that now look very different could have come from common ancestors. We can (and Darwin did) push this line of reasoning right back to the beginning of life itself: we can hypothesize that *all* living things what- soever come from a single ancestor. This would explain the nested categories. They are there for exactly the same reason that you are more like your siblings than like your cousins. All living things are one big family. Since Darwin, much further study of the patterns of similarities and differences in the light of evolu- tion, plus the fossil record, plus comparative studies of DNA, all point to the one conclusion of a single point of origin for all of life. Darwin's theory of evolution by natural selection is often recommended as being the best explanation for the apparent designedness of living things.[11] It is true that Darwin's theory has that advantage, but this is not the only reason to recommend it; the fact that it can be extrapolated from observed facts is a separate reason.

The history of Darwin's theory since his time has been one of inexorably growing acceptance by biologists and, indeed, by scientists in general. Some of the obstacles in the way of accepting it as an account of the whole of life were removed by subsequent discoveries soon after his death. Lord Kelvin (Thomson 1862) rejected Darwin's theory on the grounds that the sun and the earth could not be old enough. This was because he thought, among other things, that the sun generated heat by combustion (burning), which would have meant that it could not be more than 500 million years old. However, we now know that it generates heat by nuclear fusion, which allows it a far greater lifetime. Fleeming Jenkin (1867) rejected Darwin's theory on the grounds that an organism's traits were then thought to be a *blend* of those of its parents, so that any trait, no matter how advantageous, could not spread through a population as Darwin said it would. But in fact, even in Darwin's lifetime, Gregor Mendel had pro- duced evidence, fitting Darwin's theory, of traits being inherited in a digital way; that is, each offspring will have either the trait of its mother (e.g. blue eyes) or the trait of its father (e.g. brown eyes), but not a blend (e.g. one blue eye and one brown eye). However, Mendel's work was not well known at the time, and only became widely known at the beginning of the twentieth century.

Evidence for the gradual modification of species over long periods of time was already available in Darwin's lifetime and before from the study of fossils. This evidence has grown enormously since then. Further evidence has come from comparing the DNA of different organisms, from patterns of anatomical similarities and differences, and from vestigial (i.e. no longer functioning) organs. Darwin's inability to explain how traits are inherited, and how new variations are initially produced, began to be remedied with the rise of genetics, and has been further remedied by discoveries in developmental biology (see Chapters 2 and 5).

The theory of evolution is now absolutely central to biology, in the same way that quantum mechanics and relativity are central to physics. It is not a dispensable part of biology. There are controversies surrounding the theory, to be sure. But we should not confuse disagreements among people who accept the theory, with disagreements over whether the theory is true or not. We can divide the former type of disagreement into two sub-types:

- There are disagreements over *details* of the theory. For example, how exactly did life originate? Was the emergence of something like human intelligence inevitable once evolution got going? How important are genes in evolution and development?
- There are disagreements over what the *implications* of the theory are. For example, what, if any, implications does evolution have for human nature? For ethics? For religion?

Much of the rest of this book will be about disagreements of these two kinds.

2. The power of genes

In the 1970s, a new consensus emerged among evolutionists, inspired by the work of William Hamilton (1964a,b) and George Williams (1966), and ably defended and popularized by Dawkins in his book *The Selfish Gene* ([1976] 1989). This view retains many of the essential features of Darwin's own view as outlined in Chapter 1, and retains both the aspiration to eliminate intelligent design from the explanation of biological traits, and the claim to have success-fully done so. The orthodoxy of the 1970s has two central pillars:[1]

- *Gene selectionism (or gene centrism)*. It holds that, in so far as the traits of organisms are designed by evolution, they are designed, not necessarily so as to benefit the organism, but so as to benefit genes.
- *Adaptationism*. It sees natural selection as the primary force driving the modification of traits, and therefore as the primary tool of evolutionary explanation of traits.

The former marks a departure from Darwin's own view; the latter is claimed by some to do so.[2] Of course, what matters is not whether these views are faithful to Darwin, but whether they are true. The Hamilton–Williams view is what many people know as "the theory of evolution" from popular science and media sources. It has, apparently, become difficult for many people to conceive that there could be alternatives to this version of evolutionary theory, other than religious or crackpot-science ones. But there are, and some of those alterna-tives reject one pillar, some the other and some both. Chapters 3 and 5 of this book will be devoted to arguments for and against the first pillar, and Chapter 4 to arguments for and against the second. Here I shall present an outline of the first pillar.

2.1 Introducing the gene

The Hamilton–Williams view of evolution, and some of its alleged implications, have become famously associated with the phrase "the selfish gene" (after Dawkins's book). We have become familiar with the ideas that genes are "blueprints" or "recipes" for building organisms; that organisms "serve the interests" of their genes; that the purpose of organisms is to be "survival machines" for their genes. These formulations are clearly metaphorical: the first one speaks of organisms as if they were technological artefacts; the second imputes *interests* to genes as if they themselves were agents; and the third does both. The organism-machine analogy has its uses, and we can speak of entities – for example, animals and plants – as having "something that is good for them", and therefore purposes, even if they do not have intentions. However, from seeing organisms as (in some way) like machines it takes a further step to see them as like machines for which genes are the blueprints. From seeing organisms as having purposes it takes a further step yet to seeing genes, which are only little bits of organisms (are they not?), as having purposes. And does it not sound a bit strange to say that the purpose of a machine is to make blueprints for building that machine? If we are to make these metaphors acceptable, we need to carefully spell out:

 (i)   the comparison between a gene and an agent;
 (ii) the comparison between the relationship between a gene and an organism and that between a blueprint and a machine; and
 (iii) the comparison between the relationship between a gene and an organism and that between an agent and a machine the agent uses for its own purposes.

It sounds like a tall order, but this does not seem to have been a very strong deterrent against using "selfish gene" language. Dawkins coined many of these usages, and has on a number of occasions defended them.[3] Let us see, then, what the literal truth lying behind the metaphors is supposed to be. I shall spend a little time on the question of whether the metaphors are an acceptable way of expressing the literal truth, assuming it is true, and a lot more time on the question of whether it is in fact true.

We are familiar with claims of the kind: scientists have discovered the "gene for" *x*, where *x* might be blue-eyedness, colour-blindness, dyslexia, alcoholism or the like. It is generally agreed that there is a gene for haemophilia (the condition that Tsar Nicholas II's son Alexis inherited from his maternal grandmother Queen Victoria). What this is generally understood to mean is that the gene in question is a *difference-maker* with respect to that trait: if a creature has the gene, it has the trait; if it does not have the gene, it does not have the trait. The story is, of course, more complex than a simple biconditional ("if and only if"). In many cases there is more than one gene such that if the creature has either of them it has the trait. And a creature could have the genes "for" some trait but, because of something that happens to it during its development, does not develop the trait. For example, a person could have the genes for good, strong muscles, but could fail to develop them due to poor nutrition or lack of exercise.

In terms of actual scientific research, there are two ways in which the claim that there is a "gene for" x is made good. First, the presence or absence of a certain gene might be *reliably correlated* with the presence or absence of a certain trait. This means that, if we have studied a large sample population of some type of creature, some of which have gene G1 and some instead have gene G2, and we find that those with G1 are more likely to have trait x than those with G2, we might count that as evidence that G1 is the gene for x. This is of course a highly simplified description, as there may be more than one alternative to G1 (more than two alleles at a chromosomal locus), but let us stick to it for the sake of clarity. The creatures making up the population may vary in all kinds of ways other than whether they have G1 or G2; each has a whole set of *other* genes, which will not be uniform across the population, and the environmental conditions and life history of any two creatures will not be *exactly* the same. What a study of this kind would show, if successful, is that G1 is *correlated* with trait x, just as we might say that grey hair is correlated with old age.

Secondly, we might study the effect of a gene in artificially controlled conditions. We do our best to hold all other relevant, or potentially relevant, factors constant, and we see what happens when we change one gene. This is the idea behind the many genetic experiments that are carried out on *Drosophila* fruit flies. They are raised in laboratories, so that the environmental conditions, and what happens to them as they grow, are highly controlled. Thus, any variation that appears (some being bigger than others, say) can with reasonable confidence be attributed to genetic differences, as other kinds of difference are kept to a minimum. Moreover, the fruit flies are the products of several generations of selective breeding, which is intended to minimize differences between their overall genotypes (i.e. the total set of genes possessed by each individual fly). Having set up these conditions, experimenters can then bring about changes in specific genes, and see what happens. Ideally, in such circumstances, any variation in the flies can be attributed to the gene that has been tampered with, as other possible sources of variation can be ruled out.

Neither of these approaches is absolutely foolproof. In the case of studies on wild populations, a correlation between a variation in a gene and a variation in a trait could indicate no more than that both are correlated with some third thing, which is the actual cause of the variation in the trait. Moreover, the developmental process of any organism is highly complex and potentially vulnerable to being disrupted by any number of outside influences. Although much can be done in laboratories to reduce variation in environmental conditions, we could never be *absolutely* sure that *every* potentially relevant condition had been adequately controlled for. Examples from the subdiscipline called developmental biology give us reason to be cautious. Developmental biology studies the processes by which the initial single-celled entity that every creature starts off as (in the case of sexually reproducing organisms, a zygote) becomes an adult organism. In mammals, much of the research in the subject concerns what happens in the womb. It has been found that even very tiny alterations to the chemical environment in the womb can have catastrophic effects on how the creature turns out. The tragic case of thalidomide victims in the 1960s demonstrates this vividly (see Gilbert 2003: 694–703). The drug thalidomide brought about extremely small changes in the chemical balance of mothers' blood, which in turn led to babies being born with severely deformed arms and legs. The fact that developmental processes can be vulnerable to extremely small environmental variations means that experimenters can never be absolutely sure that all potentially relevant environmental variations have been adequately controlled for, and hence cannot be certain that any observed variation *must* be due to genetic differences. Nonetheless, some control is better than none at all, and it is not my intention here to raise sceptical doubts about the evidential value of *Drosophila*-related research. Even apart from any limitations specific to biological research, scientific findings are always provisional, as good scientists will admit. However, a number of highly impressive results have been achieved in *Drosophila* experiments. It has been said that "*Drosophila* is so popular, it would be almost impossible to list the number of things that are being done with it" (Manning 2006). For just one example, Brian Paul Lazzaro *et al.* (2006) have documented correlations between genetic differences and differences in various immunities. Given that the genome of *Drosophila melanogaster* has been fully sequenced, and that such experiments are carried out in highly controlled laboratory environments, *Drosophila* experiments provide *extremely persuasive*, if not infallible, evidence for saying that in these cases differences in genes are the cause of differences in traits. (In Chapter 5, however, we shall see a challenge to that way of seeing things.)

So we have a sense in which it is meaningful to say that Gx is the gene *for* trait *x*. Can we go further, and say that every trait has a gene for it? No. Many genes have multiple effects. For example, the *Ebony* gene in *Drosophila* causes the flies' bodies to be dark-coloured, and also causes them to be physically weak. Conversely, many traits come about through the joint effects of more than one gene. There is often no simple one-to-one correspondence between genes and traits. In any event, it is not always obvious what we should count as one trait and what as two traits, so it is not always obvious what it would mean to say that there was a one-to-one correspondence. Nonetheless, we might still think that an organism's genotype *overall* was, collectively, the genotype for the collection of traits overall that that organism has. Since that is a rather odd way to put it, we might alternatively say that if we accept, say, that gene Gb specifies that a person will have blue eyes, then the person's genotype as a whole specifies the total collection of traits that the person will have.

To illustrate the difference between saying this and saying that every trait has a gene for it, we can use an analogy by Dawkins. Dawkins rejects the oft-used turn of phrase "blueprint" to describe the relationship between genotype and organism, saying it would be better to think of the genotype as a *recipe*. In a blueprint for a machine, say, we would expect parts of the blueprint to correspond to parts of the machine: if part A is next to part B on the blueprint, then it should also be next to it in the machine. But we would not expect, for example, one part of the text of a recipe for a cake to correspond to one part of the cake.

But the claim that *every* trait is specified by a gene is clearly too strong. If I had been adopted and raised in Norway I would speak Norwegian, even though I would still have the genes I inherited from my Irish parents. To say that *if* we hold every other factor constant and vary one gene and *then* find that a trait varies, then we can attribute the variation to the gene, is surely trivial. It does not answer the question: what traits actually do vary in that way? Nor does it answer the question: when we find variations in non-controlled conditions, where a whole host of factors – some genetic, some environmental – may vary, should we attribute those variations to this gene, to that gene or to some environmental factor? It is not denied by anyone that environmental variations alone can produce variations in the outcome, or that the expression of a gene is affected by the environment. In what sense, then, are genes said to be blueprints, or recipes, for building organisms, in a way that other factors that affect those organisms are not?

Recall that I said that there were two pillars to the Hamilton–Williams view, the second being the centrality of natural selection. As I briefly discussed in Chapter 1, it is a highly salient feature of living things that they contain parts that are (*prima facie*) purposeful. Even before Darwin, it was evident that the purposes that these parts ultimately served were the survival and reproduction of the organism. Darwin developed a template for explaining how there come to be creatures with parts that are (apparently) so well designed to serve those ends: through natural selection, those that did not would be weeded out. But a key part of that story is that the traits involved must be passed down from parent to offspring. Moreover, since natural selection, on the Darwinian picture, builds up traits by very gradual accumulation over very many generations, the traits must for the most part be *very faithfully* passed down. Mutations are part of the story, but for the process to work mutations must be the exception. This is where the view that genes have a unique role to play begins to seem plausible. Those who espouse this view like to stress that genes are *unique* in how faithfully they are passed on from parent to offspring. While there is a correlation between genes and traits, no trait is passed on with the same degree of fidelity; you may have inherited genes for being tall from your mother, who in turn inherited them from her father, but you are probably not exactly the same height as your mother or her father. By contrast, the genes themselves are transmitted (mutations aside) with absolute fidelity.

But it is not just that something must be very faithfully passed down in order to be successful in natural selection; it is precisely *in virtue of* being faithfully passed down that that something is successfully naturally selected. This means that whatever that is, it must be fecund and capable of withstanding the vicissitudes of life, but crucially it also means that it must have this high copying fidelity. *Only entities that have this high copying fidelity, then, can truly be said to evolve by natural selection.* Genes, according to the Hamilton–Williams view, are the only entities that possess a sufficiently high copying fidelity. So it is genes that are the objects of evolution by natural selection. But on the Darwinian picture, evolution by natural selection is the only way in which "designed" traits – such as eyes, hearts and so on – with their complex systems of interacting parts, can come into existence. This means, then, that all these traits owe their existence to the natural selection of genes.

Two questions that arise here are "How can this be?" and "So what?" The following two sections will address these two questions respectively.

2.2 Genes and how organisms are made

I do not propose to enter into an overly technical discussion of how exactly genes produce their effects, but will say a brief word or two (with apologies to readers who already know this). Every cell in your body contains forty-six chromosomes, of which you inherited half from your father and half from your mother. Other species have different numbers of chromosomes. These chromosomes are the loci of DNA (deoxyribonucleic acid), which has a highly complex molecular structure consisting of long strands with, at their periphery, sub-molecular components called "nucleotides". There are four different kinds of nucleotide – cytosine (C), guanine (G), adenine (A) and thymine (T) – and these are the basic components of what is often referred to as the "genetic code". Yet another popular metaphor compares the four nucleotides to letters of the alphabet. Each strand of DNA thus contains a particular sequence of the four nucleotides – e.g. a very small segment of such a sequence might be ATTGCCGAATCGTTAGG – and each person (identical twins and clones apart) has an overall set that is unique to him or her.

The total ordered set of nucleotides of an individual organism is known as that individual's "genotype". The single-celled zygote that is formed from the union of the sperm and egg cells of your mother and father contains the DNA in its nucleus, from where sections of it are transcribed on to RNA (ribonucleic acid). RNA, in turn, catalyses the synthesis of proteins. Exactly what proteins are synthesized, and how much of each, is specified by nucleotide sequences: a particular section of DNA specifies a particular protein. (However, to the best of our current knowledge, much – perhaps most – DNA has no such function, and is consequently known as "junk DNA".) When the zygote splits into multiple cells, each cell contains the same unique genotype, and likewise when they in turn split. The result is that every somatic cell in your body – that is, every cell apart from your sperm or egg cells – is genetically identical. And in every cell it is the genotype that determines what proteins are synthesized, and thus the chemical makeup of the cell.

This raises the question: what then determines the differences between cells in different parts of the body? Answering this question is the task of developmental biology, and has given rise to the important subdiscipline called evo-devo, which will be dealt with in Chapter 5. The upshot of all this seems to be that the coded message that is spelt out in the alphabet of cytosine, guanine, adenine and thymine is what specifies the chemical composition of every cell in an organism, and the structure of the organism.

Proteins, and therefore all parts of the body, are made from other substances floating around in the cell, not from the genes themselves. But the claim is not that an organism is made of genes, but that genes specify the makeup of an organism. And this fits with the idea of genes being "recipes". A cake is not made of recipes, but the recipe specifies how the cake is made.[4] Equally obviously, the environment in which a creature grows up affects the way it turns out, as already noted above. This is why I have been saying that the gene *specifies* an organism's makeup, rather than determines it. Your genes do not determine how you will turn out, and even the most die-hard gene-centrist does not claim that they do. But it is claimed that they severely constrain the range within which it is possible for your development to vary: people with "genes for" tallness will not all be the same height, but they will be taller on average than people without them. (In Chapter 6, however, we shall see that things are not as simple as this.) Large effects from the environment are far more likely to produce a creature that is unviable than one that is very different. Genes thus may be thought of as having a *homeostatic* effect on traits, keeping them relatively constant from generation to generation. The idea of homeostasis is that something is kept relatively close to some state by a regulating mechanism. A thermostat provides a simple example: a certain temperature is specified by the person who sets the thermostat. When the ambient temperature goes above this, the thermostat turns down the heat; when it goes below, the thermostat turns up the heat. The temperature does not remain *absolutely* constant, but it is prevented by the thermostat from deviating far from that specified by its setting. Genes are the reason you resemble your relatives more than non-relatives, and the genes you share with all or nearly all human beings, but not animals, are the reason you resemble other human beings more than animals.

Moreover, it is the genes and not the traits themselves that are passed from parent to offspring. While the traits are variable depending on the vicissitudes of the environment, the genes are not (once again, mutations aside, but at any rate the difference in *degree* of exactitude of copying is what is alleged to be crucial). Variations that are acquired as a result of what happens during an organism's lifetime do not alter the genotype of that organism, and hence do not get passed down. This is an absolutely central dogma of modern genetics; indeed, it has been called *the* central dogma. It is the contemporary version of the rejection of the claim that (as we saw in Chapter 1) is misleadingly named "Lamarckism". This means that, in so far as traits are inherited, their inheritance is in virtue of genes being inherited. So, it is claimed, genes have a unique role in the development of traits in an organism, and in the inheritance of those traits.

2.3 Genes as agents

So now we can see the meaning of the metaphor of genes as "blueprints" for organisms. Hopefully, too, it is now clear why the discovery of DNA was an important step forwards for Darwinian theory. It relieved us of the need Darwin was under to remain silent on why organisms inherit traits from their parents. This is one reason why the discovery of DNA and the genome projects for various species, including human beings, are generally regarded as scientifically important.

However, this at best only partly explains why gene-centrist claims attract the huge amount of public media, popular science and philosophical attention they do. True, if genes really have the powers it is claimed they do, then technological manipulation of genes has far-reaching potential effects. This raises both high hopes about medical applications, and deep fears about scientists playing God, Victor Frankenstein or Doctor Moreau. But it takes more than that to explain the widespread excitement about genes.

A recurring fear surrounding gene-centrist claims is that in some way they suggest to many people that we are mere puppets with genes pulling the strings (yet another metaphor – it seems that the public perception of genetics is saturated with them). A slightly more sophisticated variant of the latter worry is that gene-centrists are not to blame if they are simply reporting the truth, but that perpetrating the claim in advance of good evidence for it is simply giving people excuses for bad behaviour (i.e. "My genes made me do it!"). To be fair, it cannot be said that gene-centrists are really guilty of making any claim as strong as the puppet metaphor suggests. At any rate, Dawkins, Steven Pinker,

Daniel Dennett and the other usual suspects often vigorously deny such a view, and none of them deny that human behaviour is highly flexible, or claim that environmental factors – such as education and parenting – make no difference. (Whether the fears, especially the sophisticated variant, are adequately answered by these denials, is something I shall return to in Chapters 13 and 14.) Moreover, if our actions are determined by our environment instead of our upbringing, it is not clear how that would make us any the less puppets. More generally, as Janet Radcliffe Richards has argued, fears about whether we are in control of our own actions can be raised in connection with *any* causal factor in those actions, and are as old as the philosophical debate about free will and determinism. The possibility that genes might be the relevant causal factor does not alter the state of play of this debate (Richards 2000: ch. 6). The fear of lack of freedom, then, is, according to the gene-centrists themselves, based on a misunderstanding of their position. (I shall say more about this issue in Chapter 13.) But the potential medical or biotechnological applications do not exhaust the reasons for those gene-centrists' claims of great public importance for our knowledge of genes.

Nor do they explain the other metaphors that are frequently employed in this discourse: those in which genes are agents and organisms are tools or vehicles designed by those agents for their own, not the organisms', purposes. There are two ideas here: first, that genes have something like *interests* that can be *served*; and secondly, that genes have something like intelligence and can design things. Surely this is anthropomorphism gone mad! Mary Midgley expresses the incredulity that many feel when faced with these metaphors: "Genes cannot be selfish or unselfish, any more than atoms can be jealous, elephants abstract or biscuits teleological" (1979: 439). How can sequences of nucleotides have intentions or intelligence? Is this not like saying that pebbles have desires? Of course, no one claims that genes actually have intentions or intelligence.[5]

However, we can sensibly talk of what is good for an animal or plant, even in the absence of intentions, and can see the parts of animals of plants as having functions that are ultimately geared towards serving that good. The beauty of Darwin's theory is that it provides a template for explaining the apparent well-designedness of parts of organisms without recourse to inbuilt purposefulness or an intelligent designer. We can quibble about whether, if we accept this theory, we should say that eyes and so on are *designed* or *serve a purpose*. Perhaps we should reserve those terms for things that have been designed by conscious, intelligent agents. Be that as it may, natural selection is alleged to give an explanation for why parts of organisms *give the appearance of* being designed and of serving purposes. (I shall leave the question of what "giving the appearance of being designed" consists in for Chapters 4 and 7.) In Darwin's own version, that explanation entails that the purpose those parts (appear to) serve is the survival and reproduction of individual organisms, and that they are designed (or appear to be) so that that purpose is served. This is because it is in virtue of their contribution to increasing those survival and reproduction prospects that those traits are selected.

The Hamilton–Williams view is different. It holds that it is in virtue of their contribution to the continued propagation of *genes* that traits are selected. This is held to follow from the consideration that for selection to act it needs something to act on that is reproduced with a very high degree of precision. Genes, it is held, are reproduced to a much higher standard of precision than any trait. This means that the parts of organisms serve the continued propagation of genes, and are designed so that that purpose is served. To take out the teleological language, the traits of organisms, in so far as they are products of natural selection, are such as to increase the chances of the genes that are involved in its construction being passed on. If natural selection is the only way that biological traits that appear to have been designed are produced, then those traits will be such as to increase the chances of genes being reproduced. And, I need hardly repeat, on the Darwinian picture natural selection *is* the only way such traits are produced.

On that Darwinian picture, natural selection is the explanation for why traits appear to have been designed for the good of the organism. For example, the fact that your heart rate speeds up in response to danger signals is useful to you. If an intelligent designer were building human bodies with the best interests of human beings in mind, that designer would probably have thought of it. We could say that Darwin's theory *predicts* that organisms will have traits that appear to have been designed to promote those organisms' survival and reproduction prospects. Suppose we were to discover some biological trait that bore the appearance of design, but did not know what it did. Darwin's theory tells us that what it does is ultimately geared to promoting the survival and reproduction of the organism to which it belongs.

We might even discover previously unknown traits in this way; if we know enough about the organism and its lifestyle in its natural habitat, we can know about threats to its survival and reproduction that it would face, and conjecture traits that would enable it to deal with those threats. Of course, this does not obviate the need to *actually look* and see whether the organism actually has the trait we have conjectured, but in principle it can help us narrow down our searches for previously undiscovered traits. In practice, such an approach has actually led to biological discoveries. Williams and Nesse (1991) list twenty-six predictions from volume 42 of the journal *Evolution* that were made on the basis of the theory of natural selection. These include that "aquatic and terrestrial snakes should differ in mass and position of ovary or young", that "distastefulness of caterpillars must have broader taxonomic distribution than gregariousness", and that "a female butterfly should prefer cryptically coloured over mimetic males". Of the twenty-six predictions, twenty were confirmed. The significance of the gene-centred view of evolution in this connection is that the explanations and predictions it generates are in terms of what serves the continued propagation of genes, rather than the survival and reproduction of individual organisms. One way to think of this, which many exponents of the Hamilton–Williams view find appealing, is that it is as if the genes were themselves deciding how they would like organisms to be, based on their selfish genic interests, and designing and programming the organisms accordingly. For example, one of the leading textbooks on evolutionary psychology has this:

> If you were a gene, what would facilitate your survival and reproduc- tion? First, you might try to ensure the well-being of the "vehicle" or body in which you reside (survival). Second, you might try directly to make many copies of yourself (direct reproduction). Third, you might want to help the survival and reproduction of vehicles that contain copies of you (inclusive fitness). Genes, of course, do not have thoughts, and none of this occurs with consciousness or intentionality. The key point is that the gene is the fundamental unit of inheritance, the unit that is passed on in the process of reproduction. Adaptations arise by the process of inclusive fitness.   (Buss 1999: 13–14)

Thus, the claim that the gene is the only thing that is truly, faithfully, inherited is thought to justify the anthropomorphic metaphors being applied to genes. It does seem that *if* that claim were true it would justify the metaphors. This is because, if the claim is true, then we should think of the purposes for which traits are designed as being *not* the well-being of the organism, but the propagation of its genes. But is the claim true? That will be subject of Chapters 3 and 5.

As the story is usually told, Darwin's theory replaced a view wherein the parts of our bodies were designed by an intelligent being who had our best interests at heart. Darwin's own theory tells us that, without any conscious designing being done, natural selection produces traits that are *as if* they were well designed with our survival and reproduction as an aim. We have other interests than survival and reproduction, but it is plausible to claim that if we were designing our own bodies, we would want them to be such that our own survival and reproduction were promoted by them. So Darwin's own theory assures us that these interests of ours are being served by the products of natural selection.

The dominant modern view, however, tells us that the products of natural selection are actually such as to promote the propagation of our genes, and only such as to promote our own survival and reproduction *in so far as that coincides* with promoting the propagation of genes. Even if the two things do actually coincide in most cases, one cannot help feeling that our being designed to survive and reproduce is only a by-product of our being designed for some- thing else. Darwin's own perspective may not itself be a very comforting one, as it reveals the "design" as illusory: the product of completely impersonal forces and no intention at all. But gene selectionism is even less comforting, as it tells us that even the as-if design that natural selection produces is not geared towards our own interests, but towards the interests of alien entities that lived long before us and will live long after us.

Despite the acknowledged success of the theory of evolution in explaining away the appearance of design in living systems, the idea that natural selection *designs* those systems appears to have a very strong, perhaps partially subliminal, grip. This, I think, is the reason why the gene-selectionist perspective is disconcerting for many people. As we saw earlier, gene selectionism does not imply that we are determined by our genes, nor does it add anything fundamentally new to any worries we might have about whether we are free. Still, it does imply that the superbly designed mechanisms in our bodies are geared to serving interests other than our own. Moreover, it implies that, *if and in so far as* we possess any innate mental mechanisms, those mechanisms are similarly geared to serving those other interests.[6] Our bodies, and at least a large part of our minds, are thus not our own servants but the servants of alien entities. The metaphors I mentioned above – of genes as agents and organisms as tools, vehicles or robots – which are so prevalent in the semi-popular polemical literature of gene selectionism, deliberately exploit the disconcertion produced by this thought, often to sensationalist effect. Here is an example from Dawkins:

> Replicators began not merely to exist, but to construct for themselves containers, vehicles for their continued existence. The replicators which survived were the ones which built *survival machines* for themselves to live in. The first survival machines probably consisted of nothing more than a protective coat. But making a living got steadily harder as new rivals arose with better and more effective survival machines. Survival machines got bigger and more elaborate, and the process was cumulative and progressive.
>
> … Four thousand million years on, what was to be the fate of the ancient replicators? They did not die out, for they are past masters of the survival arts. But do not look for them floating loose in the sea; they gave up that cavalier freedom long ago. Now they swarm in huge colonies, safe inside gigantic lumbering robots, sealed off from the outside world, communicating with it by tortuous indirect routes, manipulating it by remote control. They are in you and me; they created us, body and mind; and their preservation is the ultimate rationale for our existence. They have come a long way, those replicators. Now they go by the name of genes, and we are their survival machines.   ([1976] 1989: 19)

Douglas Hofstadter describes Dawkins's "selfish gene" view as "a topsy-turvy, disorientating, yet powerfully revealing viewpoint" (1985: 804). Disorientating it certainly is. It looks like another one of those "major blows" to the "the *naïve* self-love of men" (Freud [1917] 1963: 284) that Sigmund Freud said science had dealt. That perhaps answers the question of why it arouses so much controversy. But it does not answer the question: is it true?

3. Units of selection

The Hamilton–Williams theory of evolution focuses on genes as the "beneficiaries" of natural selection. It is alleged that genes, and genes alone, fulfil the necessary condition of having high enough copying fidelity. But natural selection is a *substrate-neutral* template, that is, it could in principle apply to any entity that fulfilled Darwin's necessary conditions of heredity, variation, population growth and limited resources. For all that I have said so far, there is nothing special about DNA that makes it alone the only thing that *could be* the locus of natural selection. Moreover, the move from Darwin's own organism-centred view to the gene-centred view of the 1970s was generally perceived to preserve a fundamental core of Darwin's theory as the same substrate-neutral template was still being employed. "Pure" gene selectionists are only claiming that *as a matter of empirical fact*, DNA is the only thing with the required high copying fidelity, not that it is the only thing that in principle could have it. Dawkins himself says that life on any other planet would have to have come about by natural selection, even though it might have a very different chemical basis from life on earth (Dawkins 1983).

Not everyone accepts the gene selectionists' claim about the uniqueness of genes. There are dissenting views over what can be said to be inherited, what it is that natural selection acts on, and what it is that is benefited when natural selection produces traits. The term that is often used for whatever these entities are is "replicators". For Dawkins, they are genes; for Darwin, they were individual organisms. The issue of what these entities are is called the *units of selection* question. Darwin's perspective, and others like it, is called *organism selectionism*; Dawkins's perspective, and others like it, is called *gene selectionism*. This chapter will deal with three other entities that have been claimed as possible units of selection: individual organisms, groups of organisms and memes.

3.1 Genes versus individual organisms

Most things that promote the survival and/or reproductive prospects of an organism also promote the replicative prospects of that organism's genes, and vice versa. For example, being able to escape predators increases your chances of living longer, and hence of having more offspring; being more attractive to the opposite sex increases your chances of having more offspring and so on.[1] But both of these mean that your genes have a greater chance of being passed on. Only a randomly selected[2] 50 per cent of your genes are passed on to each offspring, but since it is random, every increase in the number of offspring you have is an increase in the chance of any one of your genes being passed on. In terms of what is explained and predicted, then, it may seem that there is very little difference between Darwin's own view of selection – where a trait is selected in virtue of its effect on individual organisms – and the gene-selectionist view.

However, there are some crucial points of difference. Kim Sterelny and Philip Kitcher (1988) argue that the gene-centred approach to natural selection has the advantage over the organism-centred approach of *greater comprehensiveness*. That is, they claim that there are cases that can be handled by the gene-centred approach, but not by the organism-centred approach. For the requirement of greater comprehensiveness to be met, there must also be *no* cases that the gene-centred view cannot handle that the classical Darwinian view can. Sterelny and Kitcher claim that this is indeed the case. On the gene-selectionist view, genes are the primary beneficiaries of adaptation. That is, even if an adaptation happens to benefit an organism or a group, the reason it is there is that it benefits the genes. So we should not expect to find traits that benefit an organism or group at the genes' expense. If no trait that could be found that favoured either over the other – that is, if every trait that we found promoted *both* gene replication *and* the organism's survival and reproduction at the same time, if it promoted either – then there would be nothing to choose between the two views. However, there are situations where the two do conflict, and – according to Sterelny and Kitcher – in such situations we find that the genes always win. This should not be thought of as the outcome of some struggle – even in the attenuated Darwinian sense of "struggle" – between genes and organisms, which genes always win because they are so powerful. Rather it should be thought of as the outcome of the situation described above; it is in virtue of contributing to the "success" of genes that traits get naturally selected, so the traits that get naturally selected are those that contribute to the success of genes.

Yet another metaphor found in the gene-selectionist literature is of the genotype of an individual organism being a temporary alliance or cooperative venture, each gene for the most part having effects that enhance the replicative prospects of the whole genotype, rather than specifically its own replicative prospects. But this is not always the case: sometimes individual genes have effects that increase their own chances of being represented in the next generation, over and above any contribution they make to the survival and reproductive prospects of the genotype as a whole. Because they may be thought of as "defecting" from the alliance, such cases are referred to as "outlaw genes". Examples of outlaw genes include *segregation distorters* and *sex-ratio distorters*.

Segregation distorters affect the process of meiosis: that is, the splitting of a diploid (i.e. in human beings, 46-chromosome) cell into haploid (23-chromosome) cells. Normally, this process involves random "shuffling" of genes, so that the 50 per cent of the parent-cell's DNA that each haploid cell ends up with is a random selection. However, segregation distorter genes interfere with the shuffling process, with the effect of giving themselves a better than average chance of being represented in the haploid cells produced by meiosis. It is as if a certain card in a pack was somehow able to increase its own chances of coming near the top of the pack when the pack is shuffled. This effect is *independent* of any effect the gene may have on the survival and reproductive prospects of an organism containing it – it may have no effects, or it may reduce those prospects. Nonetheless, such genes will stand a better than average chance of being represented in the next generation. Thus, the persistence of such genes is a product of natural selection, but they are selected in virtue of their effect on *their own* replicative prospects, *not* their effects on the survival and reproductive prospects of the organism.

To see the significance of sex-ratio distortion genes, one must bear in mind that, if you are female, it is to your advantage for there to be as many males as possible in the population, whereas if you are male it is to your advantage for there to be as many females as possible, because this will maximize your reproductive prospects. As always in biology, there are exceptions. In this case, there are *many* exceptions, ants and bees being perhaps the most obvious. But in a great many species the ratio is 50:50, and the symmetry of the ratio is due to the interests of both sexes. However, Sterelny and Kitcher ask us to consider genes that have the effect of distorting this ratio. For example, a gene on the Y-chromosome (the chromosome that, in human beings, is present in all males but not females) might have the effect of making sperm with that gene more successful in the race to get to the egg. This – entirely hypothetical – gene has been called the "speedy-Y" gene. Since it is on the Y-chromosome only, the resulting offspring will be male. But if the speedy-Y gene became reasonably widespread that would mean that there would be more males than females in the population, to the disadvantage of the male's reproductive prospects. Nonetheless, the speedy-sperm effect of the gene could still be great enough to become widespread despite putting males in the population, including its possessors, at a disadvantage. Thus the speedy-sperm effect would be naturally selected to the benefit of the speedy-Y gene, at the expense of the organisms that possess it. Less hypothetically, evidence has been found of a gene on the

X-chromosome of *Drosophila mauritania* that reduces the production of Y-bearing sperm in males carrying the gene. There is also a Y-linked gene that counteracts this effect; if there was not, *Drosophila mauritania* would be extinct. (The sex-ratio distorting effect was only discovered when the gene was transferred to a different species, *Drosophila simulans*; Tao *et al.* 2001.)

Yet another hypothetical gene goes under the name of the "Green beard" gene. This concept was developed by Hamilton (1964a,b). As we have already seen, a single gene may have two or more effects, so there could be one that has the following three effects: (i) causing all its possessors to have a distinctive externally visible mark – say, a green beard; (ii) causing its possessors to recognize that mark in other individuals; (iii) causing all its possessors to behave in a way that benefits other individuals with green beards, and/or a way that is harmful to individuals without green beards, even to the point of self-sacrifice. "Benefit" here means something that can be reasonably expected to increase that individual's chances of surviving and reproducing. The individual making the sacrifice is not benefited, but the replicative prospects of the green beard gene are enhanced. An actual example of the green beard effect has been found in red fire ants (Keller & Ross 1998). These ants' nests have multiple egg-laying queens at any one time. There are two relevant genetic variants: *BB* and *Bb* (*bb* females die prematurely of "natural causes"). Workers can distinguish the two variants by odour (a pheromone), and those with *Bb* attack and kill queens with *BB*. However, some of the workers involved in the attacks pick up the "*BB* odour" from them, and are subsequently killed by their *Bb* fellows. Thus, *Bb* individuals benefit their own kind, even at the expense of their own lives.

Perhaps the best-known example of gene-selectionist reasoning in action is *Hamilton's rule* (Hamilton 1964a,b). You share a proportion of your genes with close relatives, over and above those that are common to the species in general. Thus, it is "in the interests" of your genes that those close relatives survive and reproduce. Hamilton devised an index of relatedness, which specifies how much benefit your genes get from a benefit to one of your relations instead of a benefit to yourself.

- For brothers, sisters, sons and daughters it is 50 per cent;
- for half-brothers, half-sisters, nieces and nephews it is 25 per cent;
- for cousins it is 12.5 per cent;
- and so on.

This means that natural selection would favour behaviour that benefited your relations, even if it was harmful to yourself, but that it tails off when it comes to more distant relatives. A sibling is "worth more" to your genes than a cousin, and so on. This mathematical index of relatedness leads us to Hamilton's rule: a behaviour that is costly to the individual is worth the cost if $C < R \times B$, that is, if the cost to the individual is less than the benefit to some other individual multiplied by how closely related that other individual is. This means that even if the cost to the individual is 100 per cent – that is, if the individual gets killed – it can still be worth it. For example, a sufficiently large benefit to three siblings – saving their lives, say – can add up to 150 per cent. Hamilton's rule, incidentally, is what makes sense of the "altruistic" behaviour of the non-reproducing workers in many species of ants and bees.

The advantage of gene selectionism with regard to Hamilton's rule is not so much that the organism-selectionist view is unable to accommodate it, but that gene selectionism is able to give a much simpler account. To accommodate Hamilton's rule on the organism-selectionist view, the formulation "survival and reproduction of the organism" has to be modified to something like "survival and reproduction of the organism, plus survival and reproduction of its relatives, in a scale determined by the genetic relatedness index". (In fact, this was how Hamilton originally understood his rule.) With gene selectionism, Hamilton's rule is accommodated by simply focusing on the replication prospects of genes, because this includes tokens of those genes that occur in other individuals, as in the "green beard" effect.

3.2 Individual organisms as units of selection

Elliott Sober (1984) argues that selection at the level of the individual organism is required to explain at least *some* features that gene selection cannot explain. His leading example is the phenomenon of *heterozygote superiority*. Recall that you inherit half your genes from your father, and half from your mother. This means that – with the exception of some genes on the X and Y chromosomes – genes come in pairs. Consider the "gene for" blue eyes, which we shall call *b*, and the "gene for" brown eyes, which we shall call *B*. Assuming that these are the only alleles that can be at this locus, there are three possible combinations a person can have: *BB*, *bb* and *Bb*. As it happens, blue eyes are a *recessive* trait, and brown eyes are *dominant*. That is, a person with *BB* or *Bb* has brown eyes, and only a person with *bb* has blue eyes.[3] As far as phenotype is concerned, whether a person has one or two *B* genes makes no difference. This dominant–recessive pattern is quite common, but it does not apply to all pairs of alleles. The gene for sickle-cell anaemia – which we shall call *s* – only gives rise to full-blown sickle-cell anaemia if a person has *ss*. But a person with *Ss* (where *S* is the normal counterpart to *s*) is not the same as a person with *SS*; a person with *Ss* is less vulnerable to malaria. This means that, if malaria is a real threat, it is better to have *Ss* than either *SS* or *ss*. In other words, the heterozygote (i.e. the one with *Ss* or *Bb*) is superior. In fact, in parts of the world where malaria has for a long time been a threat, such as Africa, sickle-cell anaemia is more common. The *s* gene keeps a foothold in virtue of heterozygote superiority. Sober argues that being *Ss* is a feature of an organism, not of any gene. Further, he says that the increase in fitness that is conferred by being *Ss* is an increase in fitness of the organism, not of any gene. Hence, according to Sober, there are some traits that are favoured by natural selection in virtue of their contribution to the fitness of individual organisms, not in virtue of their contribution to the fitness of genes.

Sterelny and Kitcher (1988) reject this argument. They point out that whenever we talk of a trait increasing the fitness of any entity, this is always *relative to some environment*. *Ss* is only an advantage to a person if that person lives somewhere where malaria is a threat. If malaria is not a threat, *Ss* is a disadvantage, as it means that some of the person's children may have sickle-cell anaemia. But the environment of a gene is not the same as the environment of an organism. Among the things that are in the environment of a gene are the other genes in whose company it finds itself from generation to generation. A gene with which it fairly often finds itself is therefore *a recurrent feature of the environment*. The times when it recurs are unpredictable, but so too is the exact time of appearance of a predator or a type of weather. This means that it is a recurring feature of the environment of the gene *s* that *s* finds itself in company with *S*. It is to recurring (not necessarily constant) features of the environment that an entity becomes adapted by natural selection. In Africa, *s* regularly finds itself in company with *S* in a situation where malaria is widespread, and in this environmental circumstance an *s* gene is at an advantage over an *S* gene. It is better, that is, to be an *s* gene than an *S* gene in this circumstance. So, Sterelny and Kitcher conclude, there is no problem accommodating cases of heterozygote superiority to the gene-selectionist perspective.

As we have already seen, Sterelny and Kitcher argue that there are things that gene selectionism can explain that individual organism selectionism cannot. So, on their view, gene selectionism wins decisively over individual organism selectionism. They call their position "pluralist genic selectionism", in acknowledgement of the fact that, in many cases, it is just as easy to describe an adaptation as being for the benefit of an organism as being for the benefit of a gene. What benefits an organism *usually* benefits its genes and vice versa. On their view, this means that it is usually just as acceptable to say that the trait was selected in virtue of enhancing the fitness of the organism, as to say that it was selected in virtue of enhancing the fitness of its genes. Either statement, they seem to be saying, is true. This is why their position is *pluralist*. The reason it is *genic* selectionism, however, is that they claim that explanations of adaptations in terms other than genetic fitness are not always available. As we have seen, they argue that not all adaptations can be explained by a contribution to the individual organism's fitness. The gene-selectionist perspective, by contrast, does not have this limitation: "The virtue of the genic point of view, on the pluralist account, is not that it alone gets the causal structure right, but that it is *always available*" (Sterelny & Kitcher 1988: 359, emphasis added). However, the arguments we have looked at so far *just* show that gene selectionism wins over individual organism selectionism in this regard. What about other possible levels of selection?

3.3 Groups of organisms, and the question of altruism

3.3.1 Why group selection?

Group selection was proposed by Darwin in *The Descent of Man* ([1871] 1981) to explain the origins of cooperation among early human beings. To put it more exactly, some behaviours appear to increase the fitness of other individuals at the expense of the fitness of the individual carrying out the behaviour. When a behaviour has this feature, it is said to be an instance of *evolutionary altruism*.

It is important to distinguish evolutionary altruism from *psychological altruism*, which is where an individual does something with the *intention* of helping another. Clearly, only creatures with a sufficiently sophisticated psychology could be psychologically altruistic: human beings and perhaps some other animals. But a tree that gave chemicals to the soil that helped other trees at its own expense would be evolutionarily altruistic.

Darwin asked how, given that the struggle for existence between individual organisms was what drove natural selection, could evolutionarily altruistic behaviours be explained. What he proposed was that such behaviours benefited the group, and hence groups of individuals that behaved in such ways would, all things being equal, grow, survive for longer and so on.

The idea of group selection has undoubted appeal, and seems consonant with the thought that natural selection is substrate-neutral. It became very popular among evolutionary theorists in the first half of the twentieth century. However, it fell out of favour in the 1960s, largely as a result of the work of Williams (1966). Williams argued that traits that favoured the group at the expense of the individual would be weeded out by natural selection. As regards evolutionarily altruistic behaviour, he argued as follows: suppose we have a group of organisms all of whom behave altruistically. It might be thought that every individual in the group would prosper as a result of this altruism, as each would benefit from the altruism of the others. And so it would be, as long as the group was *absolutely homogenous* with regard to this behaviour. However, populations of organisms are seldom absolutely homogenous with regard to any trait. Let us say that there is a group of organisms that vary with regard to some trait – how good their eyesight is, say – and that that makes a difference to the relative fitness of individual organisms: better eyesight confers an advantage. In such circumstances, one would expect natural selection to favour better eyesight so that, unless there is no selection (e.g. because resources are so abundant that pretty much every individual gets to reproduce), the ones with poorer eyesight will be weeded out. But if individual organisms in the group differ with regard to behaviour that disadvantages the behaver, that is just the same as if they differed with respect to any other selectively relevant trait.

To put it in more concrete terms, let us say that there are creatures in a population that routinely share food resources with other individuals in the population, without discrimination, and even in situations where keeping the food resources to themselves would benefit them. As long as they all do so, no one starves. However, if there are some others in the population who do not share, they will benefit from their altruistic neighbours, while not suffering the reduction in fitness that would come of sharing their own food with those neighbours. Thus, the fitness of the non-altruistic individuals will be higher, and their numbers will increase. This is so even if at the outset the non-altruists are only a very small minority. In technical parlance, the altruists are pursuing an *evolutionarily unstable strategy*, that is, a strategy that is liable to be weeded out by natural selection. In a competition between altruists and non-altruists, the non-altruists will win.

But how then do we explain apparently altruistic behaviour? To answer this, bear in mind that the altruists in the scenario just described were *indiscriminately* altruistic. Perhaps a more discriminating type of altruism could be evolutionarily stable? In fact, two such evolutionarily stable strategies have been proposed to account for apparently altruistic behaviour. The first is *kin altruism*. As we saw in Chapter 2, gene selectionism predicts that not only will natural selection favour creatures that survive and reproduce themselves, but it will also favour creatures that help out their relations. Hamilton's rule tells us that it enhances our genetic fitness to help our children and siblings, and that it enhances it less to help our nieces and nephews, and so on, according to a mathematical formula. The rule also means that making the ultimate sacrifice – giving up one's life – would enhance one's fitness if it saved the lives of three siblings and/or children, or five nieces and/or nephews and so on. But altruistic behaviour governed by Hamilton's rule would not be directed at non-relatives, and would discriminate between nearer and more distant relatives.

However, there is a second type of apparently altruistic behaviour that could be an evolutionarily stable strategy. This is what Edward O. Wilson (1978) has termed *soft-core altruism*, or *reciprocal altruism*. It might be that by helping another individual one increases the chances of being helped oneself in the future. The other individual might, for example, be an undiscriminating altruist, or might be disposed to help others who have helped her in the past. To be different from undiscriminating altruists, and to avoid being vulnerable to exploitation by non-altruists, a soft-core altruist would have to be able to distinguish between individuals who are likely to help her, and ones who are not. This could mean, for example, recognizing indiscriminate altruists, or remembering individuals who have helped one in the past. Note that this "recognizing" need not be in the full-blown psychological sense: a purely behavioural discrimination would be sufficient. There would need to be some way in which these distinctions could be made at the behavioural level, for example, some sign or mark that was "visible" to the creature and enabled it to extend help to some individuals and withhold it from others. Such soft-core altruism is extended to non-kin but, unlike kin altruism, it would not go as far as the ultimate sacrifice. There would be no point in helping another in the expectation of a future reward, if one was not going to be around to reap the reward. However, one could combine the two strategies into "soft-core inter-kin altruism" if one could be assured that those one was helping would help *one's relations* after one's own demise. This combined strategy could lead to making the ultimate sacrifice, but would still only be evolutionarily stable if it involved making discriminations.

3.3.2 Can "genuine" altruism evolve?

Nonetheless, there are still those who wish to be assured that evolution can produce a greater altruism than this, and who propose to resurrect group selection to produce that assurance. Elliott Sober and David Sloan Wilson (1998) are the most prominent of these new group selectionists. It should be emphasized that group selection is not proposed as an *alternative* to selection of genes or individual organisms. For it is not claimed, nor could it plausibly be, that all adaptations can be explained as being for the benefit of groups. Rather, those who defend group selection see it as a *supplement* to other kinds of selection. The claim is that *some* adaptations evolve that are for the benefit of groups, rather than for the benefit of genes or individual organisms, not that all adaptations are for the benefit of groups.

Sober and Wilson propose a model wherein an organism's life cycle has two stages: phase 1, where it mixes indiscriminately with a large population; and phase 2, where it confines itself to interacting within a small group. It does not matter which comes first, as long as how the creature fares in *each phase separately* affects its fitness. Let us assume that the population contains some egoists and some (indiscriminate) altruists at the outset (see Fig. 3.1). During phase 1 the egoists will do better than the altruists, as predicted by Williams. The twist in Sober and Wilson's story is that if the small groups in phase 2 are small enough, then the ratio of egoists and altruists in each group will not simply be the same as the ratio in the big group in phase 1; rather, it will vary from group to group. To understand why this is so, think of the reason that opinion pollsters like to interview *large* numbers of people: a small sample is unlikely to be representative. So it is likely that some groups will contain more egoists and some will contain more altruists, and some will have the same number of

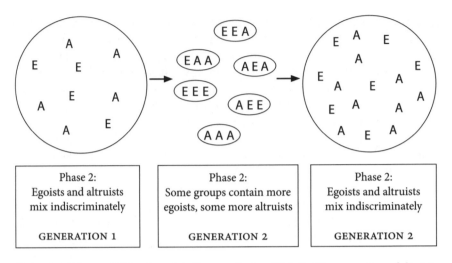

Phase 2: Egoists and altruists mix indiscriminately	Phase 2: Some groups contain more egoists, some more altruists	Phase 2: Egoists and altruists mix indiscriminately
GENERATION 1	GENERATION 2	GENERATION 2

Figure 3.1 Sober and Wilson's model of group selection. Note that the proportion of altruists has increased by phase 2 of generation 2.

each. It does not matter what the overall ratio in the big group is, as long as there are *some* altruists.

Within each small group, egoists will do better than altruists. *But* egoists in predominantly altruistic groups will do better than egoists in predominantly egoistic groups, and altruists in predominantly altruistic groups will do better than altruists in predominantly egoistic groups. In other words, *no matter what type of individual you are*, you will do better if you are in an altruistic group. That is, being in an altruistic group is fitness-enhancing. But, obviously, those in predominantly altruistic groups are mostly altruistic. If the numbers are right (i.e. the numbers representing *how much* gain in fitness you get from being helped by another), then the advantage of being in an altruistic group can outweigh the disadvantage of being an altruist, to the extent that the number of altruists steadily creeps up. Thus, the elusive result – a possible scenario in which being an altruist is an evolutionarily stable strategy – is finally achieved.

Sober and Wilson claim that their scenario is an instance of group selection, since the outcome is produced by natural selection favouring *groups* that are predominantly altruistic. This claim has been disputed by some and defended by others.[4] The situation could alternatively be described as follows: it is the feature of *being in* a predominantly altruistic group that is selectively relevant, and that is a feature of individual organisms. Moreover, if we reject individual organism-level selection, we could, at a stretch, re-describe it and say that it is in virtue of possessing a certain gene – the "gene for" altruism – that a creature is more likely to be in a predominantly altruistic group.

However, this seems to be merely terminological. It misses the point of what the group-selection controversy is really all about. It is not really about whether we should call this or that process group selection, but about whether genuine altruism can evolve. Sober and Wilson seem to have discharged the task of showing that it can. Unfortunately, when it comes to human beings, Sober and Wilson's proposal faces an empirical problem. There is no evidence that the required conditions were actually met among Stone Age human beings. Sober and Wilson's story might work for creatures that actually have this two-stage life cycle, but as far as human beings go it is simply a possible explanation for something that we do not know to exist, by means of a story that we do not know to be true.

It seems, then, that if evolution has bequeathed us innate dispositions to treat others in various ways, then they are more likely to be in discriminating ways: kin altruism or reciprocal altruism. The fact that there is a dispute over whether Sober and Wilson's scenario can be correctly described as group selection shows that, even if there are real-life examples of that scenario, they cannot be taken to have conclusively established the existence of group selection.

3.4 Memes

Although Dawkins's *The Selfish Gene* defends gene selectionism and rejects both individual organism selection and group selection, Dawkins does in fact countenance one other type of replicator that can be the locus of natural selection. This other replicator he calls a *meme* ([1976] 1989: ch. 11). Dawkins claims that the process by which *cultural* phenomena change and spread is sufficiently similar to natural selection as we know it for it too to be considered a form of natural selection. If the more familiar form of natural selection can be thought of as competition between genes, then – according to Dawkins – natural selection in the cultural sphere can be thought of as competition between memes.

What is a meme? A meme is – in a very broad sense – an idea that can be passed from one person to another by some means. Examples of memes would be: a word, a tune, a saying, a religious practice, a way of wearing one's hat, a tool, a way of cooking. If you play me a tune, and I remember it and start humming it, and then other people learn it from hearing me, and so on, then we may say that that particular meme is being replicated. Although not every instantiation of the tune is identical, there is a core that is common to all the instantiations: a sequence of notes or intervals with their note values. This core or essence of the meme can be copied any number of times without loss of information, just as a gene can be copied any number of times without loss of information. But a tune can also be modified while still remaining a "close relation" of the original.

Moreover, a modified version can end up being far more popular than the original. (The currently ubiquitous Nokia mobile phone ringtone is a slight variation of a melody by the nineteenth-century Spanish composer Francisco Tárrega.) To change the example, we might think of a *story* as having a core or essence that survives being retold in different formats, but a story might acquire some alteration and this new version might end up becoming far more popular. As they say, a tale can "grow in the telling", and many urban myths probably begin life as modifications of less spectacular real-life incidents. We can think of such modifications as "mutations", and, just as in the more familiar form of natural selection, a mutation may end up becoming the predominant form. Thus we seem to have a case for saying that the relevant features of natural selection are present: we have entities that replicate; they can and often are copied with high fidelity; sometimes new varieties are produced; and sometimes those new varieties drive out the old ones.

One might object to the idea of memes as units of selection on the grounds that they are not really analogous to genes. A gene has a physical form that is replicated, whereas at least some so-called memes do not. We may consider a tool to be a meme, and argue that all instances of a tool have a common physical form, or that a tune has the physical form of a series of waveforms. But two instances of the same slogan may be spoken or written, and moreover different spoken instances may be quite different physically (think of different accents for example), and likewise with different written instances (different handwriting, different fonts). In fact, the task of getting computers to recognize the letter "A" in many different fonts has so far proved intractable to AI researchers. Even as regards a tool, it is surely plausible to say that what makes two things instances of the same tool is that they are used for the same task, rather than that they have the same physical form. So memes cannot in general be defined by their physical properties, as genes plausibly can.

Susan Blackmore (1999: 56–8) argues that this should not be considered a problem. She argues that we did not know the mechanisms of biological inheritance prior to Crick and Watson, but that that did not stop evolutionary theory, which requires genetic inheritance, from making progress. The substrate neutrality of natural selection means that *some kind of thing* has to be copied, not that what has to be copied has to be some physical form. When we say that the same slogan may have very different physical forms, we are acknowledging that it is the *same* slogan, that is, the same sequence of words even if it not the same sequence of sounds or marks. Hence we can talk of someone "copying down" what I say. As long as we acknowledge that *something* is reproduced in a number of different instances – whether that something be a tune, a slogan, a gesture, or whatever – then the fact that it is not a physical form that is reproduced is no objection.

A further feature of Darwin's theory of evolution is that the process he describes is supposed to explain the whole gamut of apparently designed features that organisms have: the fact of the eye being well "designed" for seeing, the heart for pumping blood, and so on, are alleged to be explained by natural selection. This means that much of the structure of these features, such as the mechanisms in the iris for dilating and contracting, and hence the overall structure of organisms as a whole, is to be explained by natural selection. What, then, is the theory of memes supposed to explain? On the gene-selectionist view, what natural selection explains, and what we should expect to find, are traits that contribute to genetic fitness, that is, traits that are there in virtue of their capacity to help genes replicate. But bear in mind that memes are meant to be the replicators, so whatever the theory is meant to explain is explained in virtue of its contribution to the replication of memes.

As an example of a "memetic" explanation, Dawkins (1993) gives a feature that many religions have: the teaching that only adherents of that particular religion will be saved. According to Dawkins, this makes people more likely to believe the claims of the religion as a whole. It is *as if* the religion were designed with the express purpose of increasing its chances of being believed by a large number of people. To put it in "memetic" terms, it is as if it were designed to enhance its own prospects of replication. Just as with any natural selection explanation, we can explain this appearance of design without appeal to a conscious agent doing the design. Memes prosper because they have what it takes to prosper, not because they were designed to prosper.

An important consequence of taking the idea of selective competition between cultural entities seriously is that this competition can be *independent* of the competition that goes on between biological entities such as genes. That is, cultural entities can emerge that do not enhance genetic fitness – in fact, they may even reduce it – but that can be considered more or less fitness-enhancing in virtue of their success or failure in competition with other cultural entities. But if these entities have a fitness of their own that is not reducible to genetic fitness, then the claim that genes give a unified story, in a way that other entities do not, is gainsaid by cultural entities. And cultural traits clearly play a part in shaping many of the characteristics of human beings, characteristics that in turn contribute to the chances of the cultural traits being replicated, and hence to the chances of more human beings like the ones that currently bear the cultural traits being produced. Note too that cultural reproduction is independent of biological reproduction: one can pass on cultural traits to people who are not relatives. Priests in the middle ages, or at least those who kept their vows of celibacy, were not successful vehicles for their biological replicators, but they were highly successful vehicles for their cultural replicators.

A number of problems with meme theory might be suggested, including the following two. First, as we saw in Chapter 1, it is part of the theory of natural selection that new variations are produced *at random*, not in the sense of being uncaused, but in the sense of not being intrinsically directed towards adaptation. As we also saw, this feature is essential to what is commonly held to be the chief beauty of the theory of evolution: its ability to eliminate prior or intrinsic purposefulness from the explanation of the apparent designedness of traits. That is, the *apparent* designedness of living things can come about from processes that are not in themselves purposeful, and without any *actual* deliberate, conscious, design. Any theory of how things come to appear designed, that has them being actually designed, is simply not a Darwinian theory.

For the analogy to hold, then, we need to think of new memes as being generated in a way that is similarly random. That is, we need to think of new memes being produced in a way that does not bias them towards success. Think of a tune. If memes are analogous to genes, then for every tune that becomes a hit, there are many more that are misses. While this is probably true, it is not sufficient to eliminate prior purposefulness from the story. A person may write a tune *with the intention* of writing a hit, and it is at least plausible that those who do write tunes with this intention are more likely to have hits. George Gershwin, who wrote for Tin Pan Alley, had many more hits than Karlheinz Stockhausen, who wrote for a small clique in Darmstadt. Moreover, a person who sets out to write a hit can consciously think about what features would make her tune more commercial. Of course, she may still fail to have a hit, and another person may have a hit without intending to. But for the analogy with natural selection to hold, such conscious planning would have to *make no difference* to whether the result is a hit or not, which seems unlikely.[5] In at least some cases, the connection between writing a hit and intending to write a hit is not merely accidental. And the same is presumably true of many other of our mental products. An idea that does not catch on may be a failed attempt to produce an idea that does catch on, whereas a biological mutation is not an attempt to do anything at all.

Secondly, a further consequence of the fact that cultural entities can be consciously designed is that they can be redesigned at will. An inventor, for example, who is having difficulty getting a particular design to work, can go back to the drawing-board. That is, the inventor can start afresh on a completely different approach to the one she was following up to now. She does not have to keep modifying the design she has been working on up to now in the hope of eventually producing one that works. Natural selection, by contrast, produces its effects by continuously modifying *already existing* "designs"; it *never* goes back to the drawing-board.

It might be thought: but does anyone really ever come up with a completely new idea? Even if the inventor decides to stop pursuing a line of research that she herself was pursuing up to then, chances are that the new approach she decides to take is a development of something else that was already around, or some line of enquiry that someone else was already pursuing somewhere else. This is indeed plausible, and we can even concede that it is not just "chances are" but that this is always so. However, there remains a crucial difference: although natural selection can travel very long distances by gradual modifications, every intermediary step along the way has to be a design *that works*, that is, a viable organism. That is not to say that there cannot be failed "attempts", but that between an earlier successful attempt and a later one, there must be a series of also successful attempts. However, an inventor can gradually modify a prototype that does not work, and go through any number of prototypes that do not work to get to one that does. Even if this process started out from a previously existing design that worked, there need not be an intermediate series of designs that work. In this way, the process of human invention is less *constrained* than the process of natural selection. (The significance of constraint in evolution will be discussed in Chapter 5.)

Dawkins's meme theory is perhaps best thought of as a way of making vivid the thought that ideas can have a "life of their own". To put this thought a bit more precisely, ideas – taking this term to include tunes, stories, customs, styles of dress and more – can spread independently of anyone's intentions. Moreover, they can, without anybody designing them to do so, be as if they were cleverly designed to spread. Dawkins often uses the example of religious ideas, which, he claims, often have the very features that a clever designer would give them if that designer wanted them to become widespread.[6] Dawkins's opposition to religion is well known, and for him it is almost a defining feature of religious belief systems that they are completely unsupported by evidence. To return to an earlier example, suppose that you wanted, for some reason, to design a belief system that many people would accept without any evidence. You might include as a component of the system the claim that anyone who does not accept these beliefs is condemned to some terrible punishment. And one might add that blind faith is a virtue, and that looking for evidence is itself wrong and similarly punishable. Dawkins thinks that such components of religious belief systems are part of the explanation of their popularity. It is *as if* they had been deliberately designed to be popular, but we can explain this appearance of design without having to posit any actual, intentional designing. (I shall say more about memes and religion in Chapter 12.)

In some ways, this approach is similar to Marx's approach to explaining ideology. According to Marx, the prevailing ideology – the political ideals, moral values, ideas about human nature, and so on – will tend to be such as to favour the economically dominant class. The ideology that prevails under capitalism will be a *bourgeois* ideology, that is, it will be such as to favour the bourgeoisie, and help to keep them in power. The point of similarity between this idea and meme theory is that Marx's view is *not a conspiracy theory*. The bourgeoisie, or their friends, do not sit down and think up an ideology that will help to keep them in power. Rather, such an ideology will be such as to thrive in that particular environment. It is as if it were designed to do so, but we can explain this appearance without having to posit any intentional designing.

Leaving aside religion and politics, meme theory seems quite plausible if we think of popular cultural phenomena such as fashion or the current boom in "reality" television programmes. It seems irrational that anyone should find the clothes they wore ten years ago an embarrassment today, yet many people do. No doubt part of the explanation has to do with people's vanity, snobbery and need to feel up-to-date. These may in turn be perennial features of human nature, and hence perhaps amenable to evolutionary–psychological explanations (see Chapter 13); or perhaps sociological–historical explanations are more appropriate. Be that as it may, such factors can be considered part of the environment in which fashions evolve. The multi-billion-dollar fashion industry is testimony to how these factors are exploitable. Yet this does not mean that anyone consciously plans the next change in fashion. The specific changes may reflect shifts in the cultural *zeitgeist* too intangible and unpredictable to allow this kind of planning. Meme theory seems to provide a way of making sense of this. Fashions succeed or fail for reasons; ones that succeed are somehow more attuned to that imponderable *zeitgeist*, even though nobody designs them to be. Likewise, it is doubtful that when *Big Brother* started in the Netherlands in 1999, anyone anticipated or planned that the reality-television format was going to become so big. No conspiracy need be invoked to explain the phenomenon.

But the limitations of this are indicated by my saying that ideas *can* have a life of their own, and that there *need* be no conspiracy or intentional design for popularity. As per the first problem with meme theory above, it is surely the case that people sometimes do consciously contrive pop tunes to be hits, novels to be bestsellers and so on, and it is unlikely that such people have no idea how to do so. The fashion industry may be more consciously manipulated by its movers and shakers than I suggested above. Meme theory lets us explain popular cultural success without having to invoke conscious, intelligent design, but that does not mean that people *never* consciously, intelligently design things for popular cultural success. By contrast, apart from artificially selected traits, *no* trait of a living creature is consciously designed. With that exception, natural selection is a universal explanation for why traits of organisms bear the appearance of being well designed for success. Meme theory, then, has its uses, but should not be treated as a *universal* explanation of why cultural phenomena bear the appearance of being well designed for success. It cannot, therefore, be treated as a universal explanation for all cultural phenomena.

Neither group selection nor meme selection are put forward by their advocates as processes that on their own can give a fully comprehensive account of natural selection. Rather, they are proposed as processes that go on *as well as* gene selection (or, in Sober's case, as well as individual-organism selectionism). Even if group selectionism and meme selectionism were unproblematic, it could not plausibly be claimed that *all* biological design can be explained by either group selection or meme selection. So, while proponents of these two theories deny the gene selectionist's claim to giving a fully comprehensive account of natural selection, the theories are put forward as *supplements*, not as rivals. If we accept Sober and Wilson's model as genuinely deserving the name group selection, then group selection is possible *in principle*. That this is so is not affected by the fact that we lack any evidence that the conditions that their model requires were met among early human beings. There is, then, no reason in principle why genes need be the only units of selection. For this reason, a number of philosophers of biology are willing to countenance *multiple-level selection* (e.g. Wimsatt 1980; Brandon 1988).[7] Genes are accepted as being one of the levels at which selection takes place, and perhaps the one at which most selection takes place. But whether, and to what extent, selection takes place at other levels is an empirical question.

The case of individual organism selectionism is different because it can genuinely be seen as a rival theory, but it seems to be an unsuccessful rival. There is another present-day rival to gene selectionism, however, in the shape of developmental systems theory, which will be the subject of part of Chapter 5.

4. Panglossianism and its discontents

The Hamilton–Williams view of evolution, which I outlined in Chapter 3, sees genes as having a unique role to play in evolutionary explanation. They are taken to govern the construction of organisms, and organisms and their behaviours are taken to be designed in their interests. Views that take other things to be units of selection – for example, memes or groups – claim that at least some features of organisms are designed to serve the interests of things other than genes. However, those latter theories are still focusing on *adaptation*, that is, they are still giving answers to the question: what is *x designed for*? Where they differ is on the claim that, in so far as a trait can be said to be designed at all, it can be said to be designed for the benefit of genes.

It is compatible with either the gene-selectionist view or its rivals that very many traits are not adaptations at all. If there are traits that are not adaptations then they cannot be explained as being for the benefit of selfish genes, selfish memes, groups, or anything else. They require a different kind of explanation. But we are used to hearing evolutionary theorists offering explanations of traits in terms of how those traits promote the replication of genes and, in a minority of cases, in terms of how they promote the propagation of memes or the continuation of the group. Such explanations presuppose that the traits in question are adaptations. Is this sometimes a premature assumption? Stephen Jay Gould and Richard Lewontin famously argued that it is. But before looking at their arguments, I want to look briefly at a modern argument to the effect that natural selection is unique in its ability to explain apparently designed features of organisms.

4.1 The uniqueness of natural selection

As we have seen, Darwinian natural selection explains why the parts of organisms serve purposes. Just to get clear on terminology: a trait that is useful to an organism (or to whatever the relevant replicator is) where the fact that the organism possesses this trait is explained by natural selection, is an *adaptation*. A trait that *just happens* to be useful to an organism, but that is there for some other reason, is *adaptive*, but it is not an adaptation. But sometimes the term "adaptation" is used to refer to the *process* of production of adaptations by natural selection. Hopefully, this dual usage of the term will not cause any confusion. A trait may be adaptive without the fact that it is adaptive being the reason it is there. In such a case, the trait is explained by something else other than its being adaptive. On the other hand, if we explain a trait being there by the fact that it is adaptive, we call our explanation an *adaptive explanation*.

As I pointed out in Chapter 1, *teleology* is an all-pervasive feature of the biological world. There are many obvious examples of biological traits that are *for* something: eyes are for seeing, hearts are for pumping blood around the body and so on. Dawkins (1983) takes what he calls "functional complexity" to be *diagnostic* of life, that is, if there is functional complexity there is life. William Paley (1809) famously compared the eye to a watch. Each is a system of many interacting parts. Those parts operate to serve a specified end, and in this respect a watch is like, say, an eye or a heart. The word "complexity", if it is understood to just mean having many and diverse parts, is not sufficient to capture this feature, as Dawkins points out:

> Plenty of objects are many-parted and heterogeneous in internal structure, without being complex in the sense in which I want to use the term. Mont Blanc, for instance, consists of many kinds of different kinds of rock, all jumbled together in such a way that, if you sliced the mountain anywhere, the two portions would differ from each other in their internal construction. Mont Blanc has a heterogeneity of structure not possessed by a blancmange, but it is still not complex in the sense in which a biologist uses the term. (1986: 6–7)

The watch comparison gives us a sense of what this biologist's sense of complexity is, but it is tricky to explicitly spell it out. The organization of parts *to perform a function* is one way that we might put it. Beyond a fairly restricted limit, if the parts are moved around the function is no longer performed. I shall address the question of what, exactly, we mean by "function" in Chapter 7.

For the moment the relevant sense of complexity is intuitively clear enough for us to say: anything that is complex in this way is either living, or was designed by something living. So if we found such entities anywhere in the universe, that would be evidence of life. I shall leave aside the question of just how easy it would be for us to recognize this teleology if we saw it on an alien planet. Let us take it as reasonably plausible that we could. Dawkins argues that, if life exists anywhere in the universe, no matter how different it is from life on earth, it had to have come about by natural selection. Adaptation by natural selection, he argues, is the only possible non-"miraculous" explanation for life. In other words, it is the only explanation that does not require anything more than matter moving about according to the laws of physics, without any inbuilt purpose or intelligent design. We call such an explanation a *naturalistic* explanation. (Some people use the term "naturalism" more broadly, to mean generally consistent with a scientific worldview. But if "scientific worldview" means seeing the world as our most up-to-date science sees it, then it probably amounts to the same thing: Dawkins is claiming that natural selection alone is consistent with the rest of science.) Dawkins argues for natural selection's being the *only* naturalistic explanation for the functional complexity that all living things possess, by considering other possible explanations, as follows, and eliminating them.

(i) An inbuilt capacity for, or drive towards, increasing perfection

This is the idea that there is some intrinsic force pushing living things to become more "perfect" or more "advanced", or something of that kind. As we saw in Chapter 1, Lamarck held this view, and it marks one of the genuine differences between him and Darwin. One problem with this is that the notions of "perfection" or "advancement" are very unclear. By what standard is a giraffe more perfect, or more advanced, than a tree? But a more serious problem is that, unlike an explanation in terms of natural selection, this idea requires us to posit a force over and above the physical forces that we know to exist. So, from a naturalistic point of view it is unacceptable, as it leaves us with the problem of having to account for how there is such a force. Dawkins dismisses this view as "obviously mystical", by which I take him to mean "unacceptable from a naturalistic point of view".

(ii) Use and disuse plus inheritance of acquired characteristics

If you exercise, your muscles get bigger. If you do a lot of work with your hands, the skin on them gets harder. Perhaps other parts of the body grow by such means? And perhaps improvements produced in this way are passed down to offspring? This is what is often – misleadingly – referred to as the Lamarckian theory of evolution.

Dawkins argues that the tendency of your muscles to get bigger, and for the skin on your hands to get harder, itself needs to be explained. The normal tendency is for things to *wear out* through repeated use. For example, think of what happens to a dishcloth when you use it over and over again. Dawkins writes:

In man-made machines, parts that are subjected to wear get thinner, not thicker, for obvious reasons. Why does the skin on the feet do the opposite? Because, fundamentally, natural selection has worked in the past to ensure an adaptive, rather than a maladaptive, response to wear and tear.   (1983: 20)

Moreover, for inheritance of acquired characteristics to be able to preserve useful modifications, there would have to be some way of passing on the useful ones but not the harmful ones. For example, if you lose an eye, your children are not born one-eyed. If we were to posit some mechanism whereby useful acquired traits were passed down but harmful ones were not, we would need an explanation of how there was such a mechanism. So the same problem would arise as in (i). We would be positing a force over and above the already-known natural forces.

As it happens, it is widely believed by biologists that acquired characteristics are not inherited. But that is not the real reason why this type of explanation is problematic. Natural selection provides an explanation for how useful traits come into existence, which only requires the assumption that *some* traits are passed down to offspring, not that there is any inbuilt bias towards useful ones. Thus, even if acquired characteristics really were inherited, natural selection would still be what explained why those traits were there at all.

(iii) Direct induction by the environment

Perhaps the environment somehow shapes the organism to make it fit in? After all, organisms are said to be adapted *to the environment*. Something that might fit the description of being "inducted" by the environment is the well-known phenomenon of *imprinting*. Newborn chicks or ducklings will treat the first object they see as if it were their mother, even if that object is an experimenter's finger (McFarland 1987: 303–5). But any attempt to use this to explain designed features in general faces essentially the same problems as (ii). We would have to explain *why* an organism is able to respond to cues from its environment in certain ways. The fact that ducklings imprint is itself presumably explained by natural selection. So, once again, unless we want to posit some unknown force, the real explanatory work is being done by natural selection.

(iv) Saltationism

Maybe, at least sometimes, evolutionary novelties are produced by sudden jumps ("saltations"). As we saw, Darwin emphasized that evolution always proceeded by extremely small steps. But perhaps, sometimes, major evolutionary changes can take place in large jumps: quantum leaps as it were. Dawkins says, depending on what you mean by sudden, this is either false, or true but trivial. If by "sudden" you mean "in one or two generations", then any change, other than a very minor one, that was not a disimprovement would be so unlikely as to be for all intents and purposes impossible. It would be like throwing a million-sided dice, or like taking your television, smashing it up and shaking the parts around in a dustbin, and getting a fully operational fridge out of it. This is an instance of a general principle that I explained in Chapter 1: that there are more ways for a complex system to go wrong than for it to go right. On the other hand, if by "sudden" you mean "by the standards of the fossil record" then it is true but trivial. In fossil terms, sixty thousand years is an invisibly small distance. But this would be thousands of generations or more of any creature's existence. Reasonably significant evolutionary change could plausibly happen over this period. Dawkins emphasizes this point with a thought experiment from G. Ledyard Stebbins (1982):

> He [Stebbins] imagines a species of mouse evolving larger body size at such an imperceptibly slow rate that the differences between the means of successive generations would be utterly swamped by sampling error. Yet even at this slow rate Stebbins' mouse lineage would attain the body size of a large elephant in about 60,000 years, a time-span so short that it would be regarded as instantaneous by palaeontologists.
>
> (Dawkins 1983: 25)

Any significant evolutionary change would have to be by very small steps, and there would need to be some explanation for why those steps went cumulatively in some direction. It is true, but trivial, that the pace of evolutionary change is not always the same, so that some changes are rapid, *relative to some others*. The notion of "punctuated equilibrium", as an idea akin to saltationism that is neither impossible nor trivial, has been defended by Eldredge and Gould (1972) and Schwartz (1999). I shall return to this issue a little later in this chapter.

(v) Random evolution

Perhaps changes at the level of genes[1] can build up over centuries, having no effect for a long time, but eventually issue in some novel trait. This would allow time for a large amount of change to build up without harming the organism. The thought here is that many genes have no effect at all: the "junk DNA" that I referred to in Chapter 2. So random mutations can happen to these genes and have no effects either harmful or beneficial. The results of such changes may get spread around a population, and eventually come to be present in the majority of individuals, without any natural selection. The question that immediately arises here is: but should genetic changes that have no further effects be called *evolution*? Mitoo Kimura (1982) apparently believes they should, on the basis that changes in the frequency of genes are what evolution is really all about. However, Dawkins says that he is interested in the production of *functional complexity*. Changes in the frequency of genes *that have no effects* cannot be called developments in functional complexity.

But perhaps genes that have no effects at one time can come to have effects at a later time. Perhaps, for example, a random mutation may change a gene that has no effect into one that does. And perhaps such a random change may produce a significant, functionally complex, new trait. Unfortunately, this is subject to similar problems to the non-trivial version of (iv). It is so unlikely that it can be ruled out that random changes, unguided by anything, would either singly or cumulatively produce a new, fully functional, part. It would be like throwing a normal dice a million times looking for one particular million-digit sequence (or at least one of a very, very small set of million-digit sequences).

So, by elimination, according to Dawkins, adaptation by natural selection wins. Strictly speaking, his argument is incomplete. All he does is take a number of possible candidate explanations and argue that they do not work. Yet what he is trying to argue is that natural selection is *unique* in its ability to (non-"mysteriously", i.e. naturalistically) explain organized complexity. To establish this, it is clearly not enough to establish that certain candidates fail to fit the bill. We would need to establish first that natural selection does fit it, *and* secondly that no other candidate could. I shall take it that the former was established by Darwin's arguments, which I summarized in Chapter 1. The second is trickier, as it is hard to tell how we could rule out some future genius coming up with a new, perfectly naturalistic, explanation for organized complexity.[2] For present purposes, I shall assume that it is true. There still remain some commentators who suspect that the importance of natural selection has been exaggerated.

4.2 The accusation of "panglossianism"

In 1979, Gould and Lewontin published a paper entitled "The Spandrels of San Marco and the Panglossian Paradigm". In it, they criticized an approach to evolutionary thinking that they labelled "panglossianism" or "panadaptationism". The literary reference is to Dr Pangloss in Voltaire's *Candide*, whose guiding belief is that we live in the best of all possible worlds. Pangloss's faith in this principle is so great that he invents stories to explain the most trivial features of the world as examples of God's benevolence:

> "It is demonstrable" said he, "that things cannot be otherwise than they are; for as all things have been created for some end, they must necessarily be created for the best end. Observe, for instance, the nose is formed for spectacles, therefore we wear spectacles. The legs were

> visibly designed for stockings, therefore we wear stockings. Stones were made to be hewn, and to construct castles, therefore my lord has a magnificent castle; ....".   (Voltaire [1759] 1937: 108)

And so on. Gould and Lewontin claim that an analogous fallacy is routinely committed by evolutionary biologists in their insistence on the all-pervasiveness of adaptation. That is, Gould and Lewontin argue, there is a dominant methodology that consists of picking out some trait and then inventing an adaptive story to explain that trait:

> This programme regards natural selection as so powerful and the constraints upon it so few that direct production of adaptation through its operation becomes the primary cause of nearly all organic form, function, and behaviour. Constraints upon the pervasive power of natural selection are recognized of course ... But they are usually dismissed as unimportant or else, and more frustratingly, simply acknowledged and then not taken to heart and invoked.
>   (Gould & Lewontin 1979: 585–6)

As an analogy, they offer a case of a man-made artefact where one might be tempted to impute a purpose but it would obviously be absurd to do so. This is the now famous example of the spandrels of San Marco (St Mark's Basilica in Venice). Inside the central dome of San Marco there are four arches, supported on columns by four concave downward-pointing triangles, called spandrels. Artists used the four spandrels as an opportunity to paint triangular allegorical pictures of the four evangelists. Magnificent though these pictures are, and even though they exploit the distinctive triangular shape, Gould and Lewontin contend that it would be clearly absurd to think that housing these paintings is the *reason* the spandrels were put there. Rather, they urge, the spandrels are an inevitable by-product of the decision to mount four arches, at right angles to each other, on four columns. Thus they imported a term from architecture into the field of evolutionary theory, where it has taken on a vigorous life of its own, coming to mean a trait that is a by-product of other traits (that may themselves be adaptations) but that is not itself an adaptation. The obvious error of thinking the spandrels were put there *for* the paintings is, according to Gould and Lewontin, analogous to the error routinely made by those evolutionary theorists who assume that every trait must have an adaptive explanation.

They compare the practice of thinking up an adaptive story to explain a trait to Rudyard Kipling's *Just So Stories*. Most of Kipling's stories are about how various creatures acquired various traits, for example, "How the Camel got his Hump" and "How the Leopard got his Spots". Kipling's stories are deliberately farfetched, and the insulting connotation that many of the adaptive explanations offered by biologists are similarly farfetched is fully intended by Gould and Lewontin. They claim that it is quite easy to think up adaptive explanations for traits that we already know exist, but that the mere fact that we can think up an adaptive explanation for a trait is not evidence that the explanation is true.

If you look for an adaptive explanation for every single trait, they argue, you are acting like Dr Pangloss. They label such explanations, when they are unsubstantiated, "just-so stories". On their account, the panglossian approach to explaining a trait consists of four points:

1. If one adaptive explanation fails, try another.
2. If every adaptive explanation fails, assume that there must be one.
3. Attribute any such failure to imperfect knowledge.
4. Emphasize immediate utility and exclude other aspects of form.

(*Ibid.*: 586–7)

They sum up the strategy thus: "An organism is atomized into 'traits' and these traits are explained as structures optimally designed by natural selection for their function" (*ibid.*: 585).

Gould and Lewontin claim that if you already know that a trait exists, it is rather easy to think up *post hoc* an adaptive explanation for it. However, they argue that this should not be taken as evidence for how the trait actually came about. As an example, consider the tiny front legs of *Tyrannosaurus rex*. There has been much speculation regarding what purpose they served. For example, it has been suggested that they may have had some role in sexual stimulation. Gould and Lewontin suggest that they did not serve any purpose at all.

4.3 So what is wrong with panglossianism?

Part of the problem Gould and Lewontin are pointing to stems from what they claim is the very easiness of inventing an adaptive "just-so story". This criticism may remind you (if you know it) of Karl Popper's famous criticism of psychoanalysts and Marxists: that it is too easy to think up an explanation of the required kind. Of psychoanalysis in particular, Popper wrote: "what prevents their theories from being scientific in the sense here described is, very simply, that they do not exclude any physically possible human behaviour" (Popper 1974: 985). However, although it might sound like it, Gould and Lewontin's critique of panglossianism is not a falsificationist critique. That is, they are not just saying that panglossianism is pseudo-science because it cannot be falsified. Granted, they do write: "Since the range of adaptive stories is as wide as our minds are fertile, new stories can always be postulated. And if a story is not available, one can always plead temporary ignorance and trust that it will be forthcoming" (1979: 587). However, Gould and Lewontin do not say that there is a claim being made here – something like "all biological traits are adaptations" – that is unfalsifiable. In fact, as I quoted above, they acknowledge that "[c]onstraints upon the pervasive power of natural selection are recognized [by all biologists]". That is, they acknowledge that nobody actually makes the "all" claim. However, they say that "panglossians" *act as though* they believed the "all" claim, not considering other possible types of explanation. Moreover, they make an accusation that is just as serious as Popper's; namely, that panglossians do not bother to test their just-so stories. They also say that part of the panglossian strategy is to assume that there is an adaptive explanation even if we have failed to come up with one.

However, their main arguments against panglossianism are directed to showing *that it is false*, not that it cannot be falsified. Gould and Lewontin argue that it is absurd to assume that every trait has an adaptive explanation. They point to a number of difficulties with the assumption:

- There are many different ways in which we could subdivide an organism into traits. It would clearly be absurd to think that, whatever way we subdivide it, every single trait must be an adaptation. For example, the chin is just where your neck and your face meet, and is just a by-product of the way they grow. We do not need to look for an adaptive explanation of the chin itself.

- A trait may be a by-product of another trait. In another paper, Gould (1980) discusses the case of the panda's "thumb". Pandas have opposable "thumbs", which are modifications of their wrist-bones. These are clearly useful for eating bamboo, and so are best explained as adaptations. But pandas have similar appendages on their ankles. The likeliest explanation for the latter is that, whatever genetic modifications produced the "thumbs" also produced the "foot-thumbs" as a by-product. There is no need to explain the foot-thumbs themselves as adaptations. I shall discuss this example further in the next section.

- There are constraints on what evolution can produce. Because adaptation is a process of *modifying what's already there*, many of the solutions it produces to adaptive problems are likely to be less than ideal. For example, the prostate gland is wrapped around the urethra, with the result that men are vulnerable to the well-known effects of swollen prostate. It would clearly be better if it were somewhere else. The likeliest explanation is that the prostate and urethra system evolved out of some previously existing organs, and that the basic structure could not be undone by evolution.

But if a trait is not to be explained as an adaptation, how is it to be explained? Recall that we must distinguish between a trait that is *adaptive* – that is, useful to the organism (or its genes, or whatever) in some way – and a trait that is to be explained as an adaptation – that is, where its usefulness is the reason it is there. A trait may be adaptive without being an adaptation. An example that was known to Darwin himself was the fact that the bones in the human skull are not joined together until after birth. This makes it easier for the baby to pass through the birth canal, and it is tempting to say that this trait came about as an adaptation for that reason. But in fact, as Darwin pointed out, this same feature is present in baby kangaroos, which are tiny when they are born and hence would not have a problem passing through the birth canal anyway. It is also present in birds, which do not pass through a birth canal at all. The fact that it is present in placental mammals, although it evidently has a beneficial effect, should be explained as an inheritance from a distant ancestor, rather than as an adaptation due to that beneficial effect. This example shows that Gould and Lewontin's general point is correct: a trait that could easily be mistaken for a product of adaptation might not be. Gould and Lewontin offer a number of alternative, non-adaptive, types of explanations for traits. Throughout Gould's career he has elaborated this point.

4.2.1 Formal constraints

Towards the end of their article, Gould and Lewontin offer their alternative vision of what evolutionary biology could be. They propose reviving a nineteenth-century approach to the study of biology, wherein the main focus of study was the *Bauplänen* (German for "construction plans", "building plans" or "blueprints", singular "*Baupläne*") of organisms. The idea was that a wide variety of organisms belonging to one taxonomic group (vertebrates, say) were variants of a basic design. Reviving this approach, Gould and Lewontin think, would prevent the excesses of panglossianism by ensuring that the constraints on what evolution can produce are not forgotten.

In a much later book, *The Structure of Evolutionary Theory* (2002), Gould elaborates this idea at great length. What Gould calls the formalist programme was a strongly marked feature of continental (as opposed to Anglophone) biology in the nineteenth and early-twentieth centuries. In continental biology, attempts were made to discover the common structures that underlie both the different parts of an organism and the body plans of different organisms. For example, Goethe argued that the different parts of flowering plants – leaf, fruit, flower and so on – could be understood as variations on one basic form. He thought that the different morphologies of the various parts came about through the speeding-up or slowing-down of growth of various subcomponents of the basic form. Étienne Geoffroy Saint-Hilaire argued that all vertebrates had a common bodily structure, with parts corresponding to parts; and he extended this theory to claim deep formal resemblance between vertebrates and insects. Although Gould does not endorse Geoffroy's attempts to find structural similarities between organisms all across the animal kingdom, he thinks that the number of basic body types (*Bauplänen*) is highly restricted. He claims that the basic body plans of the different phyla have existed since the "Precambrian explosion" (a relatively sudden diversifying of life forms that he claims to find in the fossil record for about 500 million years ago), and that all new forms since have been variants of these basic types (see also Gould 2000). Gould thinks this should be considered an important factor in evolutionary explanation, side by side with adaptation.

Another type of non-adaptive explanation that can be considered a type of constraint is made in terms of *pleiotropy*, or what used to be called "correlations of growth". A genetic change may produce effects at multiple locations in the body of an organism, some of which may be adaptive and others of which may simply be by-products. An example can be found in the panda's "thumb" (Gould 1980). It is plausible to claim that the pseudo-thumbs are an adaptation for handling bamboo, the panda's main food. The most likely explanation for the similar appendages on their ankles seems to be that the genes regulating the growth of the panda's wrists also regulate the growth of its ankles:

> Repeated parts of the body are not fashioned by the action of individual genes – there is no gene "for" your thumb, another for your big toe, or a third for your pinky. Repeated parts are co-ordinated in development; selection for a change in one element causes a corresponding modification in others. It may be genetically more complex to enlarge a thumb and not to modify a big toe, than to increase both together. (*Ibid.*: 24)

4.2.2 Exaptation

Along with Elizabeth Vrba, Gould developed a concept they called "exaptation" (Gould & Vrba 1982). As an example, Gould (1991) suggests that human brains may have become larger for reasons that had nothing to do with the benefits that we later had from them: the development of language, tools and so on. An increase in brain size could have increased the computational capacities of the brain, and thus created the opportunity for language and tools to be developed, but the increase in brain size may have happened without those benefits being the reason it happened.[3] It may have come about as an adaptation for some unrelated reason, or it may simply have been a lucky accident. Either way, a trait that has an adaptive use, but originated for some other reason – either for a different adaptive use or for no adaptive use at all – they call an "exaptation".

Traits that did not originate as adaptations can be cases of the more general phenomenon of genetic drift. The undirected ticking over of genetic changes over many generations can open up new possibilities for adaptation, as well as foreclosing other possibilities. As with Kimura's view, the basis for this idea is that because many genes do not have effects, and hence cannot be subject to selection pressure from the environment, changes in their distribution will occur due to mutations and variants not being weeded out. The reason why Gould and Vrba think that this is relevant to "evolution proper" is that genes that make no difference at one time may make a difference later, so that genetic drift may make available for adaptive selection combinations that would not otherwise have occurred.

Note that it is not being claimed that drift or exaptation can provide plausible complete explanations for complex functional traits. So either the inbuilt capacities of the brain to develop language are not complex – that is, they may just be an increase in computational power – or else brains evolved the relevant functional complexity after they fortuitously became bigger and more powerful for some other reason. Traits may be *co-opted* for functions that they did not originally evolve to perform. It is widely believed that the feathers of birds originally evolved for heat insulation, and only later came to be used for flying. Clearly, the feathers, and birds' bodies more generally, have evolved much further to make them better suited for flying. But feathers are still an exaptation. The possibility of evolving for flying was opened up by their originally evolving for an unrelated purpose. What Gould and Vrba are suggesting is that a panglossian approach would fail to respect the role of these lucky accidents or co-optations in evolutionary history, and consequently fail to respect their role in explaining the form that many traits take.

4.2.3 Punctuated equilibrium

Another of Gould's alleged alternatives to adaptation as a type of evolutionary explanation is *punctuated equilibrium* (Eldredge & Gould 1972; Gould 2002: 745–1024; see also Schwartz 1999). This is one of Gould's most controversial claims. As we have seen, Dawkins argues that it signifies either something that does not exist or something that no one disputes. But this is a little unfair to Gould. He certainly does not claim that major adaptive change ever occurs in a single generation. But he does claim that the fossil record appears to reveal a picture not of constant gradual change, but of long periods of stasis *punctuated* by sudden change. "Sudden" here just means sudden by the standards of the fossil record. But, as we saw, Dawkins acknowledges that evolutionary change does not have to go at a constant speed. *Prima facie* it would appear that one can be a perfect panglossian and yet think that natural selection is not constantly producing evolutionary change. But Gould proposes mechanisms to explain why there are these relatively sudden bursts of change. For example, he suggests that much evolutionary change happens as a result of one small group of a population being isolated from the rest. In general, he argues, large populations tend to stifle evolutionary change, as novelties, even beneficial ones, will tend to be swamped in the gene pool. The emergence of evolutionary novelty via isolation of small groups is widely accepted by mainstream evolutionary biologists. The emergence of new species by this means is called "allopatric speciation". But Gould thinks that its frequency and importance has been underrated. A second way in which relatively sudden evolutionary change can occur is where catastrophes wipe out whole lineages, leaving the stage to some previously insignificant group. The most famous example of this is the sudden (in palaeontological terms, that is) disappearance of the dinosaurs, which opened up great evolutionary possibilities for the mammals. So what he is saying is more momentous than that the pace of evolution varies: it is, once again, that possibilities for new evolutionary developments can be opened up and closed down by factors that are not adaptive.

Darwin appeared to believe that macroevolutionary trends (e.g. the supplanting of dinosaurs by mammals as the predominant large land animals) should be explained by extrapolation from microevolutionary trends. That is, macroevolutionary trends should be understood as accumulations of small changes without any sudden jumps. Gould is arguing that an additional factor in macroevolution complicates this picture: punctuated equilibrium. The true significance of punctuated equilibrium is that it stresses the role of events that are unforeseeable, and therefore not adaptable to. Such events can close down certain evolutionary possibilities and open up others. Were it not for the sudden extinction of the dinosaurs, mammals might never have evolved to the position they are in now. Thus it plays a role in highlighting the role of chance, and the constraints within which adaptation has to operate.

4.2.4 Some general remarks about Gould and Lewontin

Comparison of the "spandrels" paper (Gould & Lewontin 1979) with works written by Gould and Lewontin singly shows that Gould and Lewontin have different concerns. Both sets of concerns are operative in the "spandrels" paper, but they are not entirely in harmony. Lewontin's principal concern, here and elsewhere, is with the *simplifications* that he thinks we typically find in models used in evolutionary theory. He sees such simplifications as necessary to do science at all but he thinks they may distort the public perception of science and close off certain lines of empirical enquiry prematurely. In particular, Lewontin is concerned about the dividing of living things into traits. He argues that any division of the organism into parts is artificial, and distorts the reality (Lewontin 2000a: ch. III). Even the division between the organism and its environment is, he thinks, artificial (*ibid.*: ch. I). Lewontin proposes that a more true-to-life picture would portray an organism as a system of "weakly interacting forces". Gould, on the other hand, seems far less concerned about this. He is perfectly happy to repeatedly use the phrase "relative frequency" throughout his enormous book *The Structure of Evolutionary Theory* (2002). But "relative frequency" only makes sense if we can count things up: for example, the number of traits that are adaptations and the number that are not, or the number of speciation events that are punctuated equilibria and the number that are more gradual. But this would require the kind of simplification that so vexes Lewontin.

It is worth pointing out that Gould and Lewontin are not giving a lifeline to anti-evolutionists, that is, to those people who believe that evolution cannot explain this or that. They are just saying that *adaptation* cannot explain some things. But the alternatives they propose are still naturalistic, that is, are still explanations in terms of comprehensible, natural processes. They are careful to point this out themselves: "[our view] denies that the adaptationist program (atomization plus optimizing selection on parts) can do much to explain *Bauplänen* and the transitions between them. But it does not therefore resort to a fundamentally unknown process" (Gould & Lewontin 1979: 594).

As a general summing-up, Gould and Lewontin argue that adaptation takes place against a background of constraints from past evolution, and in the context of the side-effects of random events (drift and punctuated equilibrium). They conclude that we cannot assume the perfection of adaptation that the normal, adaptationist, approach to evolution does assume. Traits need not be the products of adaptation. To show that a trait is a product of adaptation takes more than the construction *post hoc* of an adaptive "just-so story". Gould and Lewontin see themselves as offering an alternative picture of evolution that involves: (i) viewing the process of evolution in terms of modifications to a *Baupläne*, which on their view severely constrains adaptation; and (ii) an increased role for non-adaptive factors in trait-formation, such as chance, drift and purely mechanical non-adaptive processes.

4.4 A storm in a teacup?

Dennett (1987; 1995) has criticized Gould and Lewontin for, as he sees it, constructing a caricature that bears no resemblance to the actual practice of mainstream evolutionary theorists. Because Dr Pangloss was a caricature of Leibniz, Dennett suggests calling adaptationist reasoning as it is actually practised the "Leibnizian paradigm". More generally, many people (e.g. Dennett 1987; 1995; Sterelny 2001) think that Gould and Lewontin's proposed "alternative" view of evolution is not as revolutionary as they try to make it sound. All evolutionists, these critics argue, know full well that adaptations are not perfect. Dawkins, for example, devotes a chapter of *The Extended Phenotype* (1982: ch. 3) to "constraints on perfection".

However, even Dennett and Sterelny acknowledge that the paper has had a beneficial influence. When a trait is discovered, there is a tendency to look for an adaptive explanation straight away. Perhaps we should have some way of deciding whether a trait is in fact an adaptation *before* we know what purpose it serves. Otherwise, searching for an adaptive explanation might be a waste of time. Two methods have been proposed for deciding whether a trait is an adaptation:

(i) Functional complexity

If a trait is composed of a complex set of tightly integrated, interacting, parts, then it is likely to be an adaptation, rather than a by-product or a result of chance. We need to be careful in deciding what "complex" means here. We should understand it as meaning: it would require a lot of design to produce it.

A trait that was merely a structural copy of some other trait would not qualify, as it might be explained as a by-product of that other trait. The panda's foot-thumb is a complex set of tightly integrated, interacting parts. But it is a copy of the (hand-)thumb itself. The thumb is clearly an adaptation, but there is no need to invoke adaptation again to explain the foot-thumb.

One way to think of it is to imagine looking at the inner workings of a machine. One can see that it is made up of many parts working together, and reasonably surmise that it has some purpose, even if one does not know what the purpose is. A fuller account of function is required, however. This subject will be dealt with in Chapter 7.

(ii) Comparative method

Suppose one is trying to decide whether a part of organism x is an adaptation or not. One can look at organisms that are closely related to x, but have different ways of life, and therefore face different adaptive problems; and at organisms that are more distantly related to x, but have a more similar lifestyle, and therefore face more similar adaptive problems.

For example, suppose we found a trait that bats had, but did not know what it was for, or even if it was for anything. We could look at bats' nearest relatives, the *Dermoptera* (colugos), or perhaps, slightly further afield, at *Scandentia* (tree shrews). And we could also look at more distantly related creatures that share lifestyle features with bats: for example, nocturnal or cave-dwelling birds, because they face the problem of how to navigate in three dimensions in the dark. If the trait we are looking at in bats is shared by those birds but not by colugos, or tree shrews, then that is evidence that it is an adaptation to the situation that the birds and bats have in common. In fact, bats share their well-known property of navigating by echolocation with some cave-dwelling birds – oilbirds and cave swiftlets – but not with colugos or tree shrews.

This is an instance of what is called "convergent evolution", where two distantly related creatures evolve to similar outcomes to fill similar niches. Many examples of convergent evolution have been noted by comparing the mammals of Australia, which were cut off from the rest of the world from about 55 million years ago, with mammals of other parts of the world. There are, for example, marsupial "moles" and marsupial "wolves". These creatures evolved *after* Australia was cut off from the rest of the world, and are not descended from actual moles or wolves. They evolved forms similar to actual moles or wolves to fill niches that are similar to those filled by actual moles or wolves. Hence, the similarities between marsupial moles and actual moles are best explained as adaptations. Douglas Futuyma offers an example of the comparative method in action:

> [T]he testes of the chimpanzee are larger than those of the gorilla, even though the gorilla is a considerably larger animal. However, the female gorilla in estrus mates with only one male, whereas female chimpanzees often mate with several males in rapid succession. These observations suggest the hypothesis that male chimpanzees compete to fertilise the female's egg, and that a male could therefore increase his probability of success by producing large numbers of sperm, which would require large testes. Thus large testes should be expected to provide greater reproductive advantage in polygamous than in monogamous species. … Later studies showed a similar difference in testis size between polygamous and monogamous species in other taxa, such as deer and birds.   (Futuyma 1998: 359)

Neither of these methods can be absolutely decisive in determining whether a trait is an adaptation. Nor are they guaranteed to cover all cases. But they can at least *provide evidence* that something is an adaptation. Gould and Lewontin's paper urges evolutionists to first look for such evidence, rather than just assuming that a trait must be an adaptation and making up "just-so stories".

Dennett and Sterelny conclude that, while Gould and Lewontin's article was a salutary influence at the time, the claims made by Gould and Lewontin are somewhat exaggerated. Moreover, the debate between panglossians and anti-panglossians may be in one important way misconceived. It is often characterized as a debate about the *relative frequency* of adaptations versus traits that are not adaptations.[4] Yet this presupposes that we can divide traits into those that are due to adaptation and those that are not. This, however, is a confusion. For any trait is only adaptive relative to a certain set of conditions, which includes not just the environment but also the already existing makeup of the organism.[5] But this includes the outcomes of whatever non-adaptive factors played a part in forming the traits. So, if adaptationist explanation is made properly, the effects of non-adaptive factors are taken into account.

Moreover, an adaptation is always a modification of some already existing design. The very word "adaptation" indicates this. To *adapt* something is to modify it to enable it to perform some new task, or to function in some new situation: we talk of *adapting* ourselves to some new situation, for example, living in a new country; a plug *adaptor* is added to a plug to make it fit a different kind of socket; a passenger airliner might be *adapted* to serve as a cargo plane.

For these two reasons, adaptationist explanation that does not take into account non-adaptive factors is not a rival to other types of explanation, but an incoherent idea. The real conclusion of Gould and Lewontin's arguments, then, is not that evolutionists should be reluctant to employ adaptationist explanations, but that in their adaptationist explanations they should be careful to take into account the full picture of what generates an adaptive problem for an organism, which includes the makeup of the organism at the time the adaptation is held to have come about.

There is a second role, other than explanation, that adaptationist reasoning can play. It can play a role in guiding research in the discovery of previously unknown and unsuspected traits. Adaptationist reasoning (appropriately taking into consideration the background conditions as above) can be used to cut down the number of traits we might look for in an organism. Williams and Nesse (1991) have pointed out occasions where this has led to fruitful research, and this type of project is more generally defended by Dennett (1995). Futuyma's gorilla–chimpanzee example illustrates the point. [6] There may, then, be an asymmetry between adaptation and other factors in projects of discovery. Non-adaptive factors are to be taken into account as part of the background (as they should be in explanation), but it is not clear how those non-adaptive factors themselves actively generate hypotheses about what traits an organism is likely to have.

One problem with Gould and Lewontin's critique in the "spandrels" paper is that, even if it is right to point out limitations of the "panglossian" approach to evolution, the alternatives it offers do not seem particularly satisfactory. One reason for this is that their alternative vision looks a bit like a hodgepodge of suggestions for ways in which things can be explained non-adaptively. They seem to be saying that in some ways the explanation of traits involves constraint, and in some ways it involves accident. Moreover, they explicitly acknowledge that adaptation is an extremely important explanatory factor in evolution. Are they, then, just saying that adaptation is very important, but things are a bit more complicated than that? If this is all, it seems a bit vacuous. Most evolutionists agree that adaptation is not some all-powerful force unaffected by constraint or accident. There is no strict, exceptionless, law that says that all biological traits are adaptations. Because there are different ways in which we can divide up organisms into traits, such a law could not even be clearly formulated, let alone be true. There is a more moderate claim: that all traits that have functional complexity are adaptations. With their example of the bones in a baby's skull, Gould and Lewontin have pointed to what may be an exception. But, for reasons I shall discuss in Chapter 10, we should not expect to find absolutely strict, exceptionless laws in biology anyway. A law that holds for the great majority of cases is the best we can expect in biology, and is not trivial. So what is all the fuss about?

It is not clear how Gould and Lewontin's approach generates any empirical research programmes. Adaptationist reasoning has been highly successful in this regard, but the claim that not all traits are adaptations does not seem to be a good candidate for being similarly successful. This is connected to the previous point; it is because the alternative on offer is left so vague and unspecified that it is hard to see how it could generate any empirically testable claims. They do show us a non-adaptive picture of evolution in the form of the *Bauplänen* of nineteenth-century biology. But, as they admit themselves, this proved to be a failure as a research programme. Their point that adaptation is constrained in some way by the results of past evolutionary history is undoubtedly correct, and is accepted across the board by evolutionists. But what is needed is a picture of *how* these constraints operate. In Chapter 5 we shall see such a picture, in the form of a research programme that has emerged in recent decades: evo-devo.

5. The role of development

Although Darwin showed us how non-directed mechanical processes could produce design (or the appearance of design), he left a number of other related questions unanswered. He did not know what the mechanism of inheritance was, but the discovery of DNA seems to have pointed us in the right direction. But we are still left with at least one other question of comparable magnitude. Given that natural selection produces design, and DNA contains the "instructions" for building organisms, how are those instructions "read"? That is, how do we get from a single-celled zygote to an organism with billions of cells of enormously diverse structure and function? Just as Darwin provided the template for explaining how we get design, subsequent advances in developmental biology provide the template for explaining how organisms develop. How they do so will be the subject of part of this chapter. Following that I shall deal with evo-devo, a new theoretical approach in biology that explores the implications of development for evolution, and *developmental systems theory*, a recent philosophical view of biology. As we shall see, developmental systems theory takes facts about development and uses them to call into question the idea that genes contain the instructions for building an organism. Before all that, however, we shall take another brief excursus into nineteenth-century biology.

5.1 A nineteenth-century idea: recapitulation

Back in the days when people were fond of finding grand patterns in history, it was often suggested that history goes in cycles; yesterday's discarded idea will be welcomed back into the fold tomorrow. A popular twist on this view is that the revived old idea will not be exactly same as the old version, but somehow new and improved. When we think of the history of science, we generally want to think that ideas change, not just because of the vagaries of fashion, but for good, rational, reasons. Ideas, we hope, are rejected because there are reasons to reject them. This means that if a discarded idea returns, it must be in some altered form that is able to withstand the criticisms that brought the original version down.

It has been suggested (e.g. Gould 2002) that something very like this is currently happening in evolutionary biology. The new buzz-word in the air is "evo-devo", short for "evolutionary developmental biology". Gould considers that this is a medium-sized change in our thinking about evolution; he calls it "Goldilockean", because Goldilocks wanted her porridge to be not too hot and not too cold. The advent of evo-devo is, according to Gould, not a big enough change to be called a revolution, as the Copernican revolution was. But neither is it a small enough change to be called "business as usual", as (say) the addition of yet another item to our list of asteroids would be. Moreover, some of the major ideas in evo-devo bear striking resemblances to ideas in nineteenth-century biology, ideas that we thought we had long ago left behind for good reasons. It may be that these resemblances are only superficial, and it may be that the advent of evo-devo is a big enough change to be called a revolution, or a small enough change to be called business as usual. We shall consider these questions later.

In Darwin's own time, it was widely perceived that one of the best sources of evidence for common descent was embryology. As had been observed as far back as Aristotle's time, a human embryo and that of any other mammal look very similar at early stages of development. That large-headed, stooped-over form could, to an outside observer who did not already know, just as easily be destined to be a dog as a human being. Of course, we would now settle the question by taking a sample of DNA; human DNA and canine DNA are easily distinguishable. But a person in Darwin's time would not have this knowledge. In our day, a century and a half after Darwin, the thought that human beings are a type of animal is perfectly familiar to us. But it had to be fought for, and the fossil record was not the only weapon in that fight. One of the other main weapons in Darwin's day was embryology: for example, the fact that, from apparently similar beginnings, a human being or a dog can develop.

The more embryos were studied, the more strongly emerged a picture that suggested common descent. If we go back to the very earliest stages of a human being's development, we find, first, a single cell (the zygote), then a sphere of identical cells (the blastula), then the process known as "gastrulation". This is where a cavity forms on the surface of the sphere, which will eventually become the anus. It is only after this takes place that a second cavity is formed at the opposite side, which will eventually become the mouth. In some animals, the mouth comes first, and these are known as protostomes (from the Greek words for "mouth first"). The group of creatures to which human beings belong are called deuterostomes ("mouth second"). Our group, the deuterostomes, includes all chordates – so it includes all mammals, birds, reptiles, amphibians and fish – as well as echinoderms – for example, starfish – and more. The other group, the protostomes, includes arthropods – insects, arachnids and crustaceans and others – and molluscs, plus more groups. So, at this very early stage of development, the human embryo looks not only like that of another mammal, but like the embryos of a host of other creatures as well.

So we can think of the process of embryonic development as a process of progressive *differentiation*. At an early stage in development our embryos are, as it were, generalized deuterostome embryos, difficult to distinguish from the embryo of a starfish. Later they become more differentiated; we can look at them and say that they are definitely chordates, but not be able to say whether they are destined to become fish, birds or mammals. Later on again, they have jointed legs, so they can be seen to be definitely tetrapods and not fish. (Tetrapods are the group of chordates with four jointed limbs to which all mammals, birds, reptiles and amphibians belong, but not fish.) And so on until, finally, we can be sure that we are looking at a human foetus. Suppose you were watching a film of an embryo developing, and did not already know what type of creature it was (which would entail that you did not know what its DNA was, and did not have tell-tale clues like how big it was). You might say, "It's a bilaterian of some kind – could be a fly or a fish, or any number of other things for all I can tell". Then later you might say, "OK, it's a chordate, but I don't know what kind". Later still, "It's a tetrapod" and so forth. There is a *branching pattern of divergence* in embryonic development. Moreover, this pattern bears a resemblance to the taxonomic patterns of the animal kingdom. In human embryonic development, we diverge from the creatures we are taxonomically closer to later than the ones we are more distant from. We stop looking like starfish embryos before we stop looking like fish embryos, and so on. A plausible explanation for this is that we are descended from a common ancestor with those creatures, and that the taxonomic patterns are actually patterns of descent.

All of this is obvious from a twenty-first-century perspective. But a further striking fact can be found in embryology. Some of the traits found in human embryos bear a resemblance to traits found in the *adults* of other creatures. Perhaps the best-known example is the branchial arches that appear in the fourth week of human embryonic development. These are parallel ridges that cannot help but remind one of the gills of a fish. Observations like this led nineteenth-century biologists to the thought that *our individual development follows the same path as our evolutionary history*. Or, to use the preferred terminology of the day, *ontogeny recapitulates phylogeny*. This was known as the "biogenetic law" and was vigorously championed by Ernst Haeckel at the end of the nineteenth century. The thought, then, is that we begin life looking like one of our very primitive ancestors. This bit was easy to make plausible, as we are descended from single-celled organisms, and our individual lives begin as single cells. From the point of view of nineteenth-century capacities to observe, one single cell looked pretty much the same as any other. As we develop, we come to look like one of our less primitive ancestors, and we pass through the stages of fish, primitive tetrapod and so on.

Haeckel's biogenetic law is a classic example of one of those beautifully simple ideas that impose a pattern on a wide range of facts, a bit like those grand patterns in history that I alluded to earlier. Once one has been introduced to it, it becomes tempting to see confirming instances of it everywhere. Haeckel himself claimed to have found a host of facts in embryology that fit his law. Unfortunately, his idea quickly came to be discredited. Like the "grand pattern" views of history, it fell prey to the sheer complexity and diversity of facts. Even if it is true that in some ways our embryos resemble adult fish, there are many other ways in which they do not. They do not have scales, for example, and even the branchial arches are not really gills. Like the grand historical patterns, Haeckel's law can only be made to seem plausible if we ignore many facts and place great emphasis on others. One idea that was widespread in the nineteenth century was that some creatures were "higher" in evolutionary terms than others: human beings were higher than chimpanzees, which in turn were higher than other mammals, which in turn were higher than reptiles, which in turn were higher than fish, which in turn were higher than invertebrates, and so on. Evolution was often thought of as a progression from lower to higher forms. This idea is far less widely accepted today.[1] We are not, after all, descended from chimpanzees, but from a common ancestor with chimpanzees; nor are we descended from present-day fish, but from a common ancestor with them. So, even if human embryos display features in common with adult fish, that does not *in itself* mean that our embryological development is recapitulating our evolutionary history. A more serious problem, at least to a certain cast of mind, is that there seems to be no reason *why* ontogeny should recapitulate phylogeny. This is in sharp contrast to Darwin's theory of natural selection, for Darwin was, as we saw, able to give an account of why natural selection happened.

5.2 New developments in developmental biology

5.2.1 Themes and variations
Although Haeckel's law proved to be a scientific failure, there were some grains of truth in it. There are *some* similarities between human embryos and adult fish. Moreover, tetrapods almost certainly came into existence later than fish. So

the common ancestor of human beings and modern fish was in all likelihood something that looked much more like a modern fish than like human beings or any other tetrapod. This includes having gills, so we can say that, in so far as human embryos have things that look like gills, they have features that look like features possessed by our distant ancestors.

Much more important, however, is a fact that we noted in the previous section: human embryos, at an early stage in development, look very like the *embryos* of fish at an early stage in their development. Similarly, at a slightly later stage, human embryos no longer look like those of fish, but still look like those of other tetrapods. What this tells us is not that the development of human embryos follows the same path as human evolutionary history, but that in its earlier stages it follows the same path as many other creatures. Moreover, there is a pattern here: human embryos, as well as those of other chordates, look pretty much the same as those of bilateria in general, if you look at them early enough. Later, they no longer look like those of arthropods, annelids and so on, but they still look like those of chordates in general (including fish). Later still, they no longer look like those of fish, but still look those of tetrapods in general.

This fact is connected to another important feature in biology, that of *homology*. Often we find structures in common between different types of creature, especially so between ones that are closely related. For example, your arm has three different kinds of joints: a ball-and-socket joint at the shoulder, a hinge joint at the elbow and a gliding joint at the wrist. You have the same three types of joints in your leg, at the hip, knee and ankle respectively. So there is a *homology* – a common structure – between your arms and your legs. We can think of arms and legs as variations on a theme. But this theme did not originate with human beings; the same types of joints can also be found, in the same order, in the legs of other mammals, reptiles and amphibians, as well as in the wings of birds. That is, it is common to tetrapods in general. If you look at the skeleton of a frog, you cannot help but be struck by how its front legs look like miniature arms. This is so even though a leg and an arm perform different functions, and even though a wing performs a function different from either. Sometimes, parts can be similar because they serve a similar function. For example, the Australian "marsupial mole" has front claws for digging that look quite like the front claws of moles proper, even though they are no more closely related to each other than either is to mammals in general. When similar structures are present because of similar function, they are called *analogous*, as opposed to homologous. The explanation for homologous structures is *common descent*: they are legacies from a common ancestor. Thus, the homologous structures in our arms and legs, frogs' legs, birds' wings and so on are legacies from early tetrapods.

Moreover, these homologous structures also have common developmental pathways. At an early stage on the development of a tetrapod embryo, appendages begin to develop such that one could not tell just by looking at them whether they were destined to be arms or legs or wings. We could think of this as a kind of "basic tetrapod limb", from which all the various different types of tetrapod limb develop. So finding out just how the innumerable variations of the basic tetrapod limb come about is a rich field of study. The same can also be said of many other structures that can be seen to exhibit this theme-and-variations pattern. The appendages that extend from the bodies of arthropods are an extraordinarily varied lot, much more varied than tetrapod limbs. Think, for example, of insect legs, of which there is a surprisingly large variety. Or think of their wings, including, on many flies and mosquitoes, a pair of hindwings that are balloon-shaped and help with balancing. There are also antennae, the claws of lobsters, the lungs of spiders. *All* of these – legs, wings, balancers, antennae, claws, and lungs – are now accepted by biologists as being variations on a common theme, in the same way that tetrapod arms, legs and wings are variations on a common theme.

One of the things that motivated this extraordinary realization about arthropod appendages was the discovery, in 1909, of a large collection of fossils in what is known as the Burgess Shale, in British Columbia, Canada. Not only did the shale contain an extremely rich diversity of different creatures, but because of the way they were fossilized, scientists were able to study the soft parts of their bodies, not just the more usual bones, shells and so on. So scientists were provided with a detailed snapshot of creatures that lived in the Cambrian period, about 540 million years ago. (Similar finds have been made in Sirius Passet, Greenland, and Chenjiang, China.) As one would expect, a host of creatures was discovered that looked nothing like anything living today. One creature struck palaeontologists as so strange it was named *Hallucigenia*.[2] This sparked a debate about whether these creatures should be classified as belonging to present-day phyla, or as belonging to different phyla with no known modern representatives.[3] Be that as it may, some of the creatures seemed clearly to be related to today's arthropods. In particular, there were creatures called lobopods, because they had a large number of fat (lobe-like) legs. We might say that these stubby little legs were the basic arthropod limb of which the diverse legs, wings and so on are variations.

The combination of these discoveries with studies of embryonic development in insects and other arthropods led to the formulation of Williston's law: in any lineage where there are serially homologous parts, the *number* of those parts tends to decrease, while the *diversity and specialization* of different variations tends to increase. "Serially homologous" just means that there are *rows* of structures that are homologous. Because arthropods have bodies made up of segments, their appendages are examples of serially homologous parts *par excellence*. So the law is that earlier ones will tend to have many appendages that are more or less alike, whereas later ones will tend to have fewer appendages, but of more different kinds, and with more specialized functions. The evolution of wings, antennae, lungs and so on from the simple legs of lobopods is a perfect example of Williston's law. Like all "laws" in biology, it is a *ceteris paribus* law only: that is, there are many exceptions. Centipedes and millipedes, with their long rows of virtually identical legs, still flourish today. Nonetheless, the law is not vacuous, and it gives us some grounds for accepting the controversial idea that we can meaningfully speak of evolution having a *direction*.

We can also speak of this direction being echoed in the direction of development in an embryo. There is a reasonably clear sense in which ancestral forms tend to be simpler than current ones, and embryos in the early stages of development tend to be simpler in the same way. Generalized buds in an embryo develop into specialized organs. This, then, is the grain of truth in Haeckel's law. Moreover, the nineteenth-century idea of *Bauplänen* also seems to be making a comeback. The idea that many different creatures' body designs are variations on basic themes no longer seems as crazy as it did in the mid-twentieth century.

5.2.2 What can we learn from development?

What about the process of development itself? If our highly specialized and diversified bodies develop out of much simpler embryos, very similar to the embryos of a host of other creatures, how do they do so? A human embryo at an early stage looks very like a dog's embryo, and at an even earlier stage they both look like a fish's embryo, and so on. But a human embryo still grows into a human being, not a fish or a dog. Why not? If these sound like the kinds of questions a child would ask, so much the better. Much of science (and philosophy) arises from asking childlike questions.

The obvious answer is: because our *genes* are different from those of a dog or a fish. This is of course true, and it is an important *part* of the answer. A moment's reflection on some simple facts, however, will show that it cannot be the whole answer. Every cell in your body contains the same DNA, with the exceptions of sperm and egg cells. But there are many different kinds of cells, which perform many different functions. Think of the difference between a nerve cell and a white blood cell, for example. Another childlike question is: how does a cell "decide" what kind of cell it is going to be? The complete answer cannot be "its genes decide", because your nerve cells and white blood cells contain the same genes. The same conclusion can be reached by thinking about growing. The different organs in your body all grew and then stopped growing, which means that cells multiplied and then stopped multiplying. So we can ask: how do they know when to stop? We know what happens when they do not stop: that is what a cancerous tumour is.

For a long time, development was treated as a "black box" in biology. It has been known since the 1950s that DNA plays a hugely important role in shaping an organism, and after the chemical makeup of DNA was unravelled, much scientific effort went into discovering which genes were linked to which traits, and into mapping the entire genomes of human beings and other organisms. This remains one of the most significant scientific projects of all time, but it led to those childlike questions being sidelined.

A great breakthrough in answering these questions was made during François Jacob and Jacques Monod's research on the bacterium *E. coli*. This bacterium lives on the simple sugar glucose, but if glucose is not available it can take other sugars and break them down to produce it. For example, the more complex sugar lactose can be broken down into glucose and galactose. It does this using the enzyme beta-galactosidase. But this enzyme is only produced in any great quantity if lactose is present. So the question arises: how does *E. coli* "know" when to produce the enzyme? Scientists can identify which section of DNA is "for" producing beta-galactosidase, but this gene only produces the effect when lactose is present. What Jacob and Monod discovered was a protein (called the lac repressor) that chemically binds to the section of DNA that is involved in beta-galactosidase production. This prevents that section of DNA from being transcribed into RNA, and thus prevents the enzyme from being produced. When lactose is present, it binds to the lac-repressor, causing it to detach from the DNA, so the enzyme is then produced. We can think of the lac repressor as a kind of on–off switch for the beta-galactosidase gene. Its operation, and therefore the effect of the gene, depends on conditions in the cell's immediate surroundings. Thus we have an explanation for why different cells do different things, even though they contain the same DNA. Two genetically identical *E. coli* bacteria, one in the presence of lactose and one not, will do different things.

Many other on–off switches have been discovered in bacteria. Moreover, a similar type of mechanism has been discovered in multi-celled organisms, thus giving us an answer to the question of how a cell "knows" which type of cell to become. As ever, fruit flies were in the vanguard of this research. Studies of mutant flies with extra pairs of wings, or legs where their antenna should be, revealed that the mutations responsible were in specific regions of their DNA, collectively named the "homeobox". An alteration of a single codon in one of these genes gave rise to effects such as duplication of features (such as an extra pair of wings), or substitution of features (such as legs where antennae should be). It is as if the cells in certain parts of the fly's body "think" they are in a differ-ent part. The cells where the antennae are supposed to be grow into legs instead. Why does this happen? The answer was found by looking at the protein that is encoded by the homeobox genes, which is called the "homeodomain". In 1983, the geneticist Allen Loughran noticed that the homeodomain bore a striking similarity to the lac repressor in *E. coli*. It turned out that the homeodomain switches on and switches off various other genes in much the same way as the lac repressor does.[4] Thus, depending on which part of a fly a cell is in, certain genes will produce their effects and certain other ones will not. So we finally have a handle on the question of how cells "know" what type of cell to become. Depending on where it is, a specific subset of the genes in a cell will be activated, causing it to grow into a nerve cell, a white blood cell or whatever.

Homeobox genes and their associated proteins have been discovered in many other creatures, including different kinds of worms, insects, mice, cows and human beings. The task of uncovering the full developmental story of any of these creatures will doubtless be a long one. But the discovery of the homeobox and the homeodomain have made it manageable, turned it from (in Noam Chomsky's terms) a mystery into a problem. In doing so, it has given rise to a vast range of research programmes. By any reasonable standard, then, it is a major scientific achievement.

5.3 Evo-devo

One of the most surprising discoveries in genetics is the high degree of similarity between the homeodomains of different organisms. The homeodomain is made up of sixty amino acids and, of these, fifty-nine are present in flies, frogs and mice. Sean Carroll (2005: 64) calls this a "stunning" discovery, especially considering that the last common ancestor of the three lived more than 500 million years ago. The nineteenth-century idea of different organisms being variations on a common theme now looks less farfetched. We can think of the homeobox as a set of on–off switches that, in large part, produces different types of creature depending on which switches are activated where. These hugely different organisms are, to an extent, just different arrangements of the same basic biological gadgets. This means that homology is no longer just something that we can observe; it is now something for which we have at least the beginnings of an explanation. The suspicion of some scientists that homologies extended even over different phyla was unpopular even in the nineteenth century, but it now has some credibility.

This should probably not be considered a *revolution*, since the basic template of Darwinian evolution – common descent with natural selection – remains untouched. However, it does allow us to think more clearly about the implications of common descent. Its impact on biology has undoubtedly been significant. The processes that shape development are now beginning to be much better understood. Thanks to the discovery of the homeodomain, we have now solved an apparent mystery: that of how cells differentiate. This achievement is comparable to Darwin's solving of the apparent mystery of how mindless processes can produce design. We now have a much clearer idea of what factors, other than genes, shape an organism. We have discovered that many of those factors are just as complex and finely tuned as genes themselves, and just as important to the outcome.

Moreover, we can think about evolution itself in a different way. Rather than thinking of it as changes in the form and behaviour of organisms, we can think of it as changes to the developmental processes by which organisms are shaped. It is because it opens up this new perspective on evolution that it is called *evolutionary* developmental biology. Proponents of evo-devo like to point out that the study of evolution and the study of development have gone on in isolation from each other for too long. For all its faults, Haeckel's project was an attempt to unify our views of evolution and of development. Evo-devo proposes to bring them back together again after a century-long separation.

We can now meaningfully talk about *laws* in development, and in evolution, such as Williston's law. We can even talk of there being parallels between development and evolution. Even though these are only *ceteris paribus* laws, they are not vacuous. Moreover, unlike Haeckel's law, they are standing up to empirical tests, and we are able to point to reasons for them being the case.

Gould and Lewontin's talk of constraints on natural selection, and *Bauplänen*, can now be given more substance. Rather than simply saying that there are constraints, and pointing to individual examples, we now have a better sense of the "shape" of those constraints, and how they operate. The nineteenth-century interest in homologies has been revived, and rather than simply studying what homologies exist, we can understand why they are there. Evo-devo has achieved this by bringing some concepts to new prominence in evolutionary thinking.

5.3.1 Generative entrenchment

One of the concepts that has been brought to prominence is *generative entrenchment* (Schank & Wimsatt 1986). William Wimsatt (1999) explains this concept by comparing the development of an organism to the building of a house, or the development of a complex system of ideas, such as a scientific theory. The relevant thing that links these three types of things, he says, is that they all involve generative entrenchment. When you are building a house, some things have to be done before some other things. You have to lay the foundations before you build the walls, and you have to do that, in turn, before you can add the windows or the roof. The later things on this list *depend* on the earlier things being done properly. If the foundation is insecure, the walls may fall down; if the walls are not built well, the windows or roof may be affected. Once the house is built, some parts of it are easier to change than others. We can rebuild the walls without having to take apart the foundation, but we cannot re-lay the foundation without having to take apart the walls.

Note that this is *not* simply a function of the order we do things in. We could paint the walls before putting in the windows, but we would not have to take out the windows to repaint the walls. Rather, it is the fact that X *depends on Y* that makes it relatively easy to change X while leaving Y untouched, but harder to change Y while leaving X untouched. Moreover, things that are more basic features of the structure have a more significant effect on the shape of the structure. The shape of the foundation places limits on what we can build on it, but not the other way around.

Wimsatt points out that we can say similar things about biological structures. The development of an organism has a sequence of steps that usually go in a specific order. You, gentle reader, unless you are a computer or an extra-terrestrial, began life as a single-celled zygote that then split into many cells and differentiated into cells of different types. In this process the different types of cells have to turn up in the right place, and the basic structure has to be laid down, for the rest of the developmental process to take place at all. During gastrulation the cells are organized into three layers, and the body takes on bilateral symmetry, with a front and a back, a top and a bottom. [5] Clearly, if this did not happen no embryo would develop at all. Later on in development, other structures get laid down, such as the basic tetrapod skeletal structure of a spine, four limbs, ribs and a skull. The development of an organism is vastly more complex and layered than the making of any human artefact, but the same point can be made as in the case of a house. Some features are more fundamental than others, not just in the sense of being laid down earlier, but in the sense that if they did not take place, the others would be severely affected, but not vice versa. Moreover, more fundamental structures limit the possibilities for the final product. We have a tetrapod skeletal structure in common with other mammals, birds, reptiles and amphibians. The similarities are quite extensive; for example, the pattern of the types of joints you have in your arms (ball-and-socket joint at the shoulder, hinge joint at the elbow, and gliding joint at the wrist) is also seen in your legs, and in the legs, wings or what have you of all tetrapods, with the exception of those that have "lost" their limbs, such as snakes. Although this structure has been adapted to many different tasks, evolution is *conservative* in that it has not replaced that structure for hundreds of millions of years. The snakes and other exceptions have *lost* their limbs; they have not replaced them with a different structure. This is not just because adaptation is slow; it is because adaptation, as I said in Chapter 3, is *modification* of something that is already there. It is harder to modify structures that many other things are affected by than ones that they are not.

This should remind you of Gould and Lewontin's *Bauplänen*. The word, remember, literally means "construction plans", "building plans" or "blueprints", and so suggests an analogy between the building of a house and the development of an organism. As we saw, we may think of the wings of birds, the arms of human beings and the legs of frogs as variations on a theme (homologues). The reason homologues exist is that evolution is conservative, in the sense just outlined. Features that are hard to change without having to redesign all the other things that are "built on top of them", are said to be "generatively entrenched". By means of this concept, we can use our knowledge of how organisms actually develop to gain a much clearer picture of constraints on what evolution is likely to produce. We shall return to the concept of generative entrenchment in Chapter 6.

5.3.2 Modularity

A second concept that has risen to prominence as a result of evo-devo is *modularity*. The key idea here is that we can think of an organism as composed of modules. A module can be thought of a subsystem that is, to a degree, self-contained. It is a *sub*system, in that it is part of a larger system, as your heart is part of your body. But it is self-contained to the extent that the parts of the system interact with each other more than they interact with other parts of the system. Note that this is only "to a degree". There is *some* interaction between a module and other parts of the system; if there were not, the module would not be part of the system at all. The idea is illustrated in Figure 5.1, which depicts two modules. The circles represent components of each module, and the arrows represent interactions. The two boxes are modules, because there is more interaction between the parts within each box than there is between the two boxes.

Why should we think that organisms are structured in this way? There are both empirical and theoretical reasons. We can see some very clear empirical examples of modularity in nature. A simple case is the segmented bodies of earthworms. Many of an earthworm's organs – such as seminal vesicles and nephridia (similar to kidneys) – are confined within the walls of one segment, rather than crossing the segment boundaries. The theoretical reasons for believing in modularity stem from an understanding of how natural selection actually works.

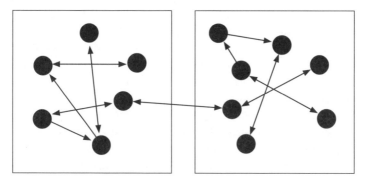

Figure 5.1 Two modules and their interactions.

A key idea in the first of these arguments is that adaptations arise in response to *specific* problems in *specific* contexts. At one time, there may be a problem of avoiding a particular predator. Such a problem might be dealt with by, for example, equipping the creature with camouflage. But different predators might have different sensory apparatus. For example, a creature that was red–green colour blind would fail to see things other creatures see, and also see things that other creatures fail to see. Thus, if a creature develops camouflage, it had better be camouflage that works against the particular creature it needs to avoid. It may, in addition, make it harder for other creatures to see it as well, as a by-product. But there is no such thing as a general-purpose camouflage: one that is effective against all creatures. The point is that selection pressure is generated by specific circumstances, and only produces adaptations to deal with those specific problems. The greater the selection pressure, the more finely tuned we would expect the adaptations to be. And this implies dedication to specific tasks. This point is made by Donald Symons: "There is no such thing as a 'general problem solver' because there is no such thing as a general problem" (1992: 142).

Another evolutionary argument points to the advantage of mechanisms that are *de-coupled* from each other. The thought here is that further selection pressure might modify a mechanism, and it can do so with less disruption of other mechanisms if the different mechanisms are relatively de-coupled. A creature could, for example, evolve to be more sensitive to certain colour distinctions, without other mental functions being disrupted as a by-product (Brandon 2005).

The study of development has given us a greater understanding of more reasons why we find modularity in living things. Early in development, organisms' bodies tend to divide up into regions. After a zygote is formed, it grows and divides into a cluster of identical cells. But during gastrulation, those cells start to differentiate into different kinds, by the kind of process involving homeobox genes that I mentioned above. The growing embryo also starts to divide very early into different regions. Which region a cell is in will determine which genes are "switched on" in that cell. These regions behave like the modules depicted in Figure 5.1; there may be some interaction between the regions, but there is usually much more interaction between the parts within each region. As development progresses, finer distinctions into regions occur. We can think of the limb buds of tetrapod embryos as modules, for example. Thus, the division of organisms' bodies into modules tends to happen early in development, and thus to affect how the organisms eventually turn out.

The modules in an embryo from these early stages of differentiation onwards are thus modular in two ways: there is more interaction between the parts within a module than between different modules, as in Figure 5.1; and there are genes that are relatively specific in what modules they affect, as in Figure 5.2. Note that this need only be *relatively* specific, as Figure 5.2 indicates: some genes can affect more than one module. There is a very important difference

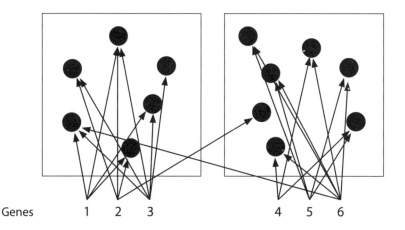

Figure 5.2 Although there is not a one-to-one gene–trait relationship, a group of genes has far more influence on one group of traits than another.

of degree here. I pointed out in Chapter 2 that the "one gene, one trait" view is far too simple. But that does not mean that we can say no more than that the whole organism is affected by the whole genotype. The effects of any particular gene are far greater in some regions than in others. While development is going on, the parts of organisms interact with each other, and this takes place in a modular way. A diagram depicting the interactions that affect the development of an organism would have to *combine* Figures 5.1 and 5.2, as interactions of both types go on.

Developmental modularity is closely related to generative entrenchment, in that organisms develop by progressively finer differentiation into modular regions. The earlier a division of this kind takes place, the more it is likely to affect subsequent development. As with generative entrenchment, the *mere fact* of something developing earlier does not mean that many other things depend on it; rather, what matters is how deeply entrenched it is. As with generative entrenchment, studying the ways in which organisms actually develop has revealed general principles underlying development in the widest possible range of living things. Using our knowledge of these principles, we can once again give a more definite shape to the factors that Gould and Lewontin see as constraining the directions in which evolution can go.

Someone committed to the gene-centred view of evolution that I discussed in Chapters 2 and 3 need not be troubled by evo-devo. Although we have discovered that genes' effects depend on protein switches, those protein switches – the homeodomain – are themselves coded for by genes. So it is still possible to think of genes as being in the driving-seat of both development and evolution. A different way of looking at things, arising from within philosophy, urges us that thinking about development should lead us to stop thinking of genes as "special" in this way.

5.4 Developmental systems theory

As we saw in Chapter 2, the idea that there are genes "for" certain traits has a wide appeal. This idea may be seen as a contemporary version of the idea that certain traits are "innate". We are inclined to use the expressions "there is a gene for x" and "x is innate" interchangeably. The claim that there is a "gene for" something often engenders a feeling of fatalism – if it is innate there is nothing that can be done about it – despite the disclaimers of gene-centrists to the effect that their position is not deterministic. Much of the appeal of gene-centrism, as well as the anxiety, may stem from the lingering old-fashioned idea that certain traits are innate and certain others are acquired. I shall deal at greater length with the issue of innateness in Chapter 6. For the moment, I will have to say a little about it in order to introduce developmental systems theory. Developmental systems theorists claim that a proper understanding of how organisms develop should lead us to abandon the old-fashioned idea of innateness, and that, in turn, should lead us to abandon the perspective of gene-centrism. To see how their argument for this goes, we shall look at how developmental systems theory originated.

5.4.1 Lehrmann versus Lorenz

Isolation experiments
Developmental systems theory originated with Daniel Lehrmann's (1953) critique of Konrad Lorenz's isolation experiments (e.g. Lorenz 1950). Lorenz is famous for, among other things, being a pioneer of cognitive ethology. He set out to discover which animal behaviours were innate, and in order to do this, he designed experiments where animals were raised in isolation from their normal environmental developmental cues, and observed them to see which behaviours manifested themselves as normal in such artificial circumstances. He concluded, seemingly reasonably, that, since the animals did not have the opportunity to acquire the behaviours from their environment, those behaviours must be innate. For example, rats raised in isolation, when they are coming near to giving birth, will gather up material and use it to build nests, even when it is their first pregnancy. They cannot have learned this behaviour from previous experience, or from observing other rats doing it, so, Lorenz concluded, it must be an innate behaviour.

However, Lehrmann showed that this behaviour still required specific types of interaction with an environment in order to develop. In particular, the rats had had experience of carrying food pellets, and of moving their own droppings out of the way. Lehrmann set up an experiment whereby the rats' food was powdered, and the rats were raised in cages that had mesh floors, so that the droppings fell through. In this new situation, they no longer built nests; the material they might have used was left lying around. The moral of this story is that sometimes the environmental interactions needed to develop a trait are far from obvious.

This moral was reinforced by a further finding by Lehrmann: he discovered that rats, by licking their own bodies during pregnancy, obtained potassium salts. But this behaviour also seemed to be required for them to acquire the behaviour of licking the bodies of their young. Rats who had been raised with special collars that prevented them from licking themselves did not lick their young either. So the emergence of a trait in an isolation experiment would only show that whatever environmental resources are needed to develop that trait are present in that particular artificial set-up.

Lehrmann argued that rather than dividing traits into two types, innate and acquired, one should ask of any trait which resources were necessary for it to develop. In the case of the rat developing these behaviours, the relevant resources do not include seeing the example of other rats doing the same thing, but do include prior exposure to objects that can be picked up and carried around. The fact that these are the relevant resources is no reason to say that the traits are not "learned", if learned is supposed to mean "as opposed to instinctive", and no reason to call the traits innate as opposed to acquired. Nothing is gained by assigning these traits to the categories "innate" or "acquired", or "learned" or "instinctive".

The "learned versus instinctive" dichotomy has been entrenched in thinking about animal behaviour at least since the nineteenth century. It is encapsulated in William James's famous definition of instinct, for example: "Instinct *is usually defined as the faculty of acting in such a way as to produce certain ends, without foresight of those ends, and without previous education in the performance*" (James [1890] 1981: vol. 2, 1006, original emphasis). According to Lehrmann, such definitions serve only to obscure the need to look at the *specific developmental process* by which a trait comes about, and tempt one to set up the dichotomy between innate and acquired. He argues that we should not bother asking the question "Which traits are innate?" or its modern version "Which traits are the product of genes?" Both questions are meaningless.

Lorenz (1965) returned to the fray with an attempt to account for innateness in terms of information contained in the genes. According to such an account – which can also be seen in popular form in Dawkins's comparison of genes to a recipe – the way an organism will turn out can, under ideal conditions, be "read off" its genotype. "Ideal conditions" here does not just mean conditions in which we know exactly which genes do what. It also means situations where the environmental conditions are such that they can be treated as a neutral background against which genes do their work. Lorenz accepted the obvious point that input from the environment was necessary for any trait to develop at all:

> No biologist in his right senses will forget that the blueprint contained in the genome requires innumerable environmental factors in order to be realised ... During his individual growth the male stickleback may need water of sufficient oxygen content, copepods for food, light, detailed pictures on his retina, and millions of other conditions in order to enable him, as an adult, to respond selectively to the red belly of a rival. Whatever wonders phenogeny can perform, however, it cannot extract from these factors information that simply is not contained in them, namely the information that a rival is red underneath.
>
> (*Ibid.*: 37)

Lorenz considered that environmental factors should be treated as "channel conditions" and the genotype as a "signal". The analogy here is with, say, a radio, where if it is working correctly the signal – that is, the music or the talk or whatever – will come through clearly. Crackles and pops due to a malfunction of the radio or a problem with the transmission are interferences due to the channel conditions, rather than part of the signal.

However, as developmental systems theorists such as Susan Oyama (1985) have argued, the problem here is that deciding what counts as signal and what counts as channel conditions depends on what you are interested in. If you are a radio engineer trying to fix the radio, then you may be interested in the clicks and pops, as they may enable you to discover what is wrong with it, and whether you have fixed it. You will probably tune the radio to static so that the sounds from the transmitted programmes do not interfere with the information you are interested in. In this case, what the radio receives – that is, static, or nothing – is part of the channel conditions, and the crackles and pops are part of the signal. Likewise, if we hold environmental conditions constant in order to see what happens when we manipulate the genome, we can consider the environment to be channel conditions. But we might alternatively hold the genome constant to see what happens when we manipulate some aspect of the environment, in which case we can consider the genome to be part of the channel conditions.

Lehrmann argued that any trait whatsoever requires environmental conditions to be a certain way in order to develop, and will develop differently or not at all if the environmental conditions are varied in certain ways. This point can be seen to be trivially true, boringly obvious, if one considers that in the absence of food or oxygen, no trait would develop at all. But the idea that some traits are innate and some are acquired pervades our biological thinking, so much so that even a great scientist such as Lorenz accepted it.

Abandoning the innate/acquired dichotomy
Developmental systems theory rejects the distinction between the innate and acquired. Paul Griffiths (2002) has recently argued that this distinction belongs to folk biology, that is, it is a pre-scientific idea, like the pre-Darwinian idea of biological species as fixed and immutable kinds. Rather than asking whether a trait is innate or acquired, developmental systems theorists urge that we should ask: what is its developmental history? And what resources does it need in order to develop? Every textbook on genetics, evolution or evolutionary psychology nowadays carefully points out that the expression of a gene depends on the environment. But developmental systems theorists wonder whether such statements are merely paying lip-service to the idea, or whether the implications of this have been truly taken to heart. The idea that genes are what control the process of development retains a powerful grip on our imaginations, and it is this grip that developmental systems theory seeks to break.

We saw in Chapter 2 how it is an integral part of the Hamilton–Williams view that genes contain the information for building an organism. It is admitted by all biologists, including those who are inspired by Hamilton and Williams, that the actual development of an organism depends not just on the genes, but also requires that the right environmental and other extra-genetic resources be present. A creature that does not eat will die, no matter what information its genes contain. Moreover, it is obvious to everyone and universally acknowledged that in developing in the womb a mammal needs the resources the womb provides to develop. To say, as Lehrmann does, that environmental resources make a big difference, hardly seems to impugn the notion that genes contain the information. A cake could not be made without the eggs and flour and so on, and a building could not be built without the bricks and mortar and so on, but that does not impugn the idea that it is the recipe, or the architect's plan, that contains the information determining how the ingredients are put together. We can accept the validity of Lehrmann's experimental findings and say that what the genes provide – or, as we might otherwise put it, what is innate – is the propensity to respond to a certain environmental circumstance in a certain way. The person who has genes for being tall, say, will not be tall in just any set of environmental circumstances, but will, given the right diet and so on, be tall. All biologists accept this. Indeed, it has become standard practice in perfectly orthodox textbooks on evolution and evolutionary psychology to insist that the distinction between innate and acquired is meaningless. For example:

> As with all interactions, the product simply cannot be sensibly analyzed
> into separate genetically determined and environmentally determined

components or degrees of influence. For this reason, *everything*, from the most delicate nuance of Richard Stauss's last performance of Beethoven's Fifth Symphony to the presence of calcium salts in his bones at birth, is totally and to the same extent genetically and environmentally code-termined. (Tooby & Cosmides 1992: 83–4)

Here is another example (from a recent textbook on "biopsychology"):

Like earlier versions of the nature–nurture question, the how-much-of-it-is-genetic-and-how-much-of-it-is-the-result-of-experience version is fundamentally flawed. The problem is that it is based on the premise that genetic factors and experiential factors combine in an additive fashion – that a behavioural capacity, such as intelligence, is created through the combination or mixture of so many parts of genetics and so many parts of experience, rather than through the interaction of genetics and experience. Once you learn more about how genetic factors and experience interact, you will better appreciate the folly of this assumption.

(Pinel 2000: 24)

This is despite the fact that most evolutionary psychologists are perfectly orthodox adherents of the Hamilton–Williams view of evolution. This view is sometimes referred to as the "interactionist consensus", and is ably defended by Matt Ridley in *Nature via Nurture* (2003b). What, then, it might be asked, is so distinctive about developmental systems theory?

5.4.2 What makes genes the replicators?

As we saw in Chapter 3, the Hamilton–Williams view incorporates the idea that genes are, as it were, the real beneficiaries of natural selection. Their grounds for holding this view are that genes have a high degree of copying fidelity, and that the gene-selectionist view gives a comprehensive account of all adaptations. Developmental systems theorists challenge both of these claims.

Copying fidelity

Developmental systems theorists hold that the role of genes and environment in development is *symmetrical*. Genes are information-rich, there is no doubt, but so too is the environment. "Environment" is normally thought of as what is "outside the skin" of an organism, and when we picture the relationship between an organism and its environment we tend to think of a creature – usually in its adult form – interacting with physical entities outside itself. We think of the weather, or the local vegetation, as part of an organism's environment. Moreover, we tend to think of the environment as something we can describe quite independently of the organism: the rain is the same whether there is a creature being made wet by it or not. However, even the gene-selectionist view requires us to revise this pre-reflective concept of the environment: recall that in Sterelny and Kitcher's defence of gene selectionism, it was argued that we can see the environment of a gene as including the other genes with which it finds itself in company. Moreover, when a mammal is in the womb, we can think of the womb as its environment, providing it with nutrition, temperature conditions and other factors that shape its development. We already noted how this process of development can be disrupted by very small alterations to that environment (the thalidomide case mentioned in §2.1); what is truly remarkable is the high degree of predictability that the developmental process nonetheless exhibits. Developmental systems theorists urge that this predictability is *not* solely due to the regulating influence of genes, although that is one factor. Rather, it is due to the environment in which a creature develops being as a whole highly predictable. When we think of a womb, this should be clear enough. But developmental systems theorists urge that it is also true of what we more conventionally think of as the environment: the "world outside". They do not deny that the physical world can be an unpredictable place. But they say – and this is one of their most surprising claims – that not everything in the physical world, or even in a creature's immediate surroundings, counts as part of its environment.

Only some parts of the surrounding world are relevant to a creature, and which parts are and which are not depends on the creature's constitution and way of life. Lewontin (1983; 2000a: ch. 2) makes this point very forcefully, and argues that two different types of creature living in the same physical surroundings would have different environments. As we saw in Chapter 1, it is precisely the ever-present and ever-recurring features of the surrounding physical world that an organism adapts to. Something that only appears very sporadically and irregularly in an organism's surroundings is unlikely to become a resource for the organism; and nor, if it is harmful, is it something for which natural selection can prepare the creature. So the very possibility of natural selection presupposes that there are predictable features in the organism's surroundings. Organisms are designed by natural selection to employ reliably recurring features of their surroundings as resources in development, survival and reproduction. Lewontin's point (or one of his points) is that it is only a subset of the reliably recurring features of the surroundings that are relevant to a creature and hence are part of its environment. We are familiar with many cases where an organism, or a population of organisms collectively, shapes aspects of its surroundings. The most obvious cases of this are where a creature actively manipulates matter to make things, for example, termites building nests or beavers building dams.

But other, less obvious, processes, according to Lewontin, count as shaping the environment. Mammals and birds regulate their own body temperature, but their bodies give off heat, so that in regulating their own body temperature they are regulating that of their immediate physical surroundings as well. Trees form a canopy beneath which light and temperature conditions are significantly different from what they would be were the trees not there. In fact, Lewontin urges, it is impossible for a creature *not* to affect its surroundings, as there is a constant exchange of matter and energy (e.g. breathing, eating, excreting, giving off heat, giving off water) going on. But the fact that these processes, and life cycles as a whole, are recurring and predictable, reinforces the recurrence and predictability of relevant features of the environment. We should not say, then, that only *some* creatures shape their environment. As a further development of this thought, we can think of an organism as *part of its own environment*. Lehrmann's rats obtaining potassium salts by licking themselves is only a particularly striking illustration of this. A further way in which organisms may be said to influence the environment they are in is by "tracking" external conditions by moving about with, for example, weather conditions, thus making for themselves a situation in which their surroundings are relatively more constant than if they stayed still.

Processes by which organisms shape their surroundings are often referred to as "niche construction", and are recognized in perfectly orthodox gene-centred biology. Indeed, Dawkins devoted a whole book – which he considers his most important work – to such processes (Dawkins 1982). He urges us to see the effects produced by an organism on its surroundings as the *extended phenotype*. That is, just as the organism's morphology and behaviour are shaped by, and for the benefit of, its genes, so are the effects it produces on its surroundings. But developmental systems theorists hold that the high copying fidelity of genes is equalled by the high reliability of environmental conditions. The first purported reason for holding the gene-centred view is that the high copying fidelity of genes accounts for the highly stable reproduction of traits that is necessary for natural selection. However, developmental systems theorists emphasize that a whole host of resources, other than genes, that affect development are similarly passed down from generation to generation with high fidelity. A creature inherits a faithfully reproduced set of genes from its parent(s), but likewise it inherits a faithfully reproduced set of environmental resources, and *both* are needed to account for an organism reliably turning out to be like its parent(s).

Genes are passed down from parent to offspring in a very precise form, but the environmental niche in which an organism finds itself is not an arbitrary, chance matter; environmental niches are highly structured and an organism in the wild reliably finds itself in the same niche as its parents. That is, the *same* environmental resources are available to the offspring as to the parent, and these environmental resources shape development in a systematic way. The environmental niche of a species is just as distinctive of the species as the genome. The environmental niche can be said to *belong to* the species, just as much as the genome.

We might be inclined to think that *Homo sapiens* is a major counter-example to this, since we have the ability to flourish in many different environments. Do we not flourish in many different climates all over the earth, and have we not sent people to outer space and to the depths of the ocean? Three points may be offered in reply to this, however:

- We need to be careful to avoid the species chauvinism that would see the range of environments in which human beings live as broad in an absolute sense. We could not live under the bark of a tree, or in the intestinal tract of a sheep, although these are environments in which some organisms flourish.
- To the extent that human beings do get about in what for us are abnormal environments, we do so by creating micro-niches that are human-friendly, that is, that approximate our natural environment with regard to, for example, temperature and air pressure. (Think of a deep-sea diving suit, or the pressurized cabin of an aeroplane.)
- The early stages of our development, the stages that most shape how the organism turns out, take place in a highly predictable environment: that of the mother's womb.

We are used to thinking of the environment as influencing the expression of genes, but developmental systems theory holds that to see it in this way underplays the active role of the environment in development; it would be just as acceptable, developmental systems theorists urge, to see the genes as influencing the expression of an environmental niche. This way of describing things strikes us immediately as strained and artificial, but developmental systems theorists want us to see the more orthodox description – where genes are seen as containing all the information for constructing an organism – as equally strained and artificial. We should learn to see the sum total of the resources involved in an organism's development – the genes *and* the environment – as *collectively* containing the information for constructing that organism.

Among the factors other than genes that may affect the way an organism turns out are: the chemical constitution of the maternal egg cell; the symbiont micro-organisms that the growing foetus receives through interaction with its mother's bloodstream; the climate and presence of different foodstuffs in the organism's environment; the presence or absence of appropriate external cues for eliciting various behaviours and instigating various developmental programmes. Any of these factors is just as capable as a gene of being a *difference maker* in the development of some trait. For example, the cell membranes of an organism are initially grown directly from those in the zygote, which in turn come from those in the maternal egg cell. Genes are involved in synthesizing the proteins with which the new cell membranes are grown, but consider the symbiont micro-organisms that a foetus receives from its mother. These make crucial differences to the life of the organism, for example, there are symbiont

85

bacteria in human beings that play important roles in digestion. A variation in these micro-organisms could confer an advantage on the host organism, and thus get itself passed on to future generations. We could think of this as an adaptation of the host organism, but not of the host organism's genes. And there are resources external to the organism that affect its development in ways that, in turn, promote the propagation of those very same external resources.

Thus, developmental systems theory urges us to count all the internal and external resources that go into making up an organism as making up a highly integrated system that reproduces itself and evolves. Genes play a part in this process of course, but the part they play is not fundamentally different from that played by any number of other developmental resources. This system cannot be equated with the organism, because it includes things that are external to the organism.

Explanatory comprehensiveness

As we saw at the end of Chapter 2 and the beginning of Chapter 3, the second reason gene selectionists claim that their view is superior to others is that it is able to accommodate the full range of biological adaptations. That is, according to gene selectionists all adaptations are such as to promote the replication of genes, whereas not all of them are such as to promote the replication of anything else. This claim goes hand in hand with the claim that information contained in the genes specifies (although it does not *determine*) the way an organism will turn out. It is in virtue of genes' alleged ability to regulate how organisms develop that natural selection is held to produce traits that promote the replication of genes. However, developmental systems theorists challenge the claim that genes have a unique role in regulating development. In consequence of this, they deny that gene selectionism is able to accommodate the full range of biological adaptations.

The idea that genes regulate development has much plausibility: if you have the gene for blue eyes, then you will have blue eyes whether you grow up in a hot climate or a cold climate, and whether you are a vegetarian or a meat-eater. As I explained in Chapter 2, the idea is that genes hold traits in homeostasis, securing them against the unpredictable vicissitudes of the environment, and preventing them from deviating too far from a central norm over the generations. However, on developmental systems theorists' view, genes are not solely responsible either for securing traits from environmental vicissitudes, or for keeping them in homeostasis over generations. As we have seen, gene selectionists explicitly acknowledge that the way any particular organism actually turns out is a joint product of genes and environment. But for the idea that genes act as regulators holding things in homeostasis to be given substance, it must be that it is only the *deviations* from the norm that are explained by the effects of the environment, or, for that matter, by anything other than the genes.

This brings us to the question: what do developmental systems theorists think is the unit of selection? The answer is: the *life cycle*: "the fundamental pattern of explanation – the development of complex form through variation and differential replication – is preserved. ... Evolution occurs because there are variations during the replication of life cycles, and some variants are more successful than others" (Griffiths & Gray 1994: 298).

Developmental systems theorists claim that their view is more comprehensive than the gene-selectionist or organism-selectionist alternatives:

> [D]evelopmental systems theory maximises the explanatory power of evolution. It allows the formulation in a single explanatory framework of all natural-historical narratives that are genuinely explanatory.
>
> (*Ibid.*: 288)

> [W]hen a feature is replicated, it is due to the replication of the whole process for which it is a resource. Conceiving evolution as the differential replication of developmental processes/life cycles therefore gives us maximum explanatory power, allowing us to explain everything that can be explained in terms of differential replication. (*Ibid.*: 304)

The Hamilton–Williams view, as we saw, holds that natural selection designs organisms for the benefit of their genes. Developmental systems theorists reject this view. Since, on their view, the life cycle as a whole is replicated, it follows that natural selection designs organisms for the benefit – that is, the continuance – of the life cycle as a whole. Developmental systems theory is a *holistic* view of evolution.

Against the claims of developmental systems theory, a gene selectionist might point to the success of gene selectionism *in actual scientific practice*. As the countless studies on *Drosophila* show, there is much to be discovered by examining and manipulating variations in genes. Scientists can link many genetic variations to variations in visible traits, for example, the ebony gene in *Drosophila* is linked to the distinctive dark-coloured body and physical weakness. Even though these observations are made in highly controlled laboratory conditions, they are not as vacuous as Lehrmann's critique of Lorenz might suggest. Many traits have been found to be linked to genes in ways that are robust against the actual environmental variations that an organism is likely to meet in the natural world. That is, even if the developmental systems theorists are right that there are *some* environmental conditions in which the trait will not develop, a creature possessing the gene is almost certain to develop the trait in any conditions it is likely to encounter in its natural habitat. The most obvious examples are genetically inherited defects, such as haemophilia. But this can also be the case with functional traits as well, if those traits are *canalized* (see §6.5). Such findings are by no means worthless; they may, for example, have important applications in medicine.

By contrast, developmental systems theory has not yielded much in the way of empirical findings or even research projects. This is not just for the extraneous reason that it is not very well known to scientists. Rather, it seems to be because of developmental systems theorists' insistence on the holistic nature of inheritance and development. To take developmental systems theorists at their word, we would have to control for every possible variation in the life cycle of an organism, on the grounds that it *might* make a difference to the outcome. Surely this is an utterly impracticable task. In this regard, developmental systems theory also stands in contrast with evo-devo, which is proving extremely fruitful in terms of generating research projects.

Having said that, Lehrmann's criticisms of Lorenz, and the arguments of developmental systems theorists themselves, help to relieve us of any tendencies we might have to take a fatalistic attitude towards our genes. Oyama illustrates this with a medical example:

> Take a child with phenylketonuria (PKU), a metabolic disorder that usually leads to mental retardation if untreated but that can often be controlled by instituting a special diet early in life. The normality exhibited by this child on a proper diet is not an environmentally produced normality that phenocopies genetic normality; it is the result of a particular combination of unusual genome and unusual (for us) environment. The mental retardation of a PKU child on an unregulated diet is similarly the result of coaction, or constructivist interaction; it is no more or less genetic than it is environmental. (2000: 37)

But, as we saw above, seeing traits as results of "coaction" is common among perfectly orthodox Hamilton–Williams evolutionists. Lehrmann's arguments force us to think more carefully about the concept of *innateness*. He was right to insist that genes themselves cannot determine how an organism turns out. But such notions as "innate" or "product of genes" – which we often think of as interchangeable – seem intuitively clear in their meaning. When we examine them more closely, as developmental systems theory encourages us to do, they seem to dissolve. Can they be salvaged? That will be the subject of Chapter 6.

6. Nature and nurture

This chapter will deal with the familiar biological concept of *innateness*. We are used to hearing claims that such-and-such a trait is innate or, as it might otherwise be expressed, inborn, in our genes, part of our evolutionary heritage, and so on. We seem to know what is meant when such claims are made. Moreover, there is a long history of such claims being made and being contested; for example, Locke and Leibniz disagreed on whether there were "innate ideas" in the mind. Very often, in such debates, it seems to be just assumed that we know what "innate" means. But do we? That will be the main subject of this chapter.

6.1 Why does innateness seem to matter so much?

First I want to say a little about why the issue of whether something is innate or not often seems to be a very urgent one. Why does it matter so much whether traits are innate or not? There is often a strong political dimension to such debates. In recent years, the "human nature wars" have attracted much publicity: the battles, that is, over whether certain human traits are innate or whatever the opposite of innate is – acquired, socially constructed, learned or some such. These battles tend to become especially heated around the issue of male–female differences. Evolutionary psychologists often claim that there are typical behavioural differences between men and women, and that these differences are part of our evolutionary heritage. Others believe that, if there are such differences at all, they are products of upbringing, social conditioning and so on. Moreover, certain regrettable (at least to many people) tendencies in people are often attributed to "human nature", for example, the tendency for societies to be hierarchical, or the tendency to go to war. Again, there is vehement opposition to such claims. I shall say more about these debates in Chapter 13.

One reason why many debates about such claims tend to have a political dimension is that calling a trait innate seems to carry an implication of *fatalism*: it is just part of human nature, it has always been with us, and always will be, and so on. Fatalistic conclusions might in turn breed *quietism*: an attitude of "we might as well just accept it; there's no point in trying to change it; *que sera sera!*" As one would expect, such a conclusion is often unwelcome to those of certain political persuasions. Many (but by no means all) feminists claim that male–female behavioural differences are largely socially constructed, and consequently that we are not simply fated to live with them forever. The inference seems to be that if this is true, then there is hope for anyone who believes that people should be free to choose their own lifestyle regardless of their sex. The further thought seems to be that liberation movements would be futile if people were naturally destined for certain gender roles anyway. Likewise, if we could show that war, or social inequality, are not inevitable given human nature, it would give encouragement to pacifists and egalitarians. A further consequence of this is that people, depending on their political convictions, may have a vested interest in claiming that something is or is not innate. Marxists have for many years insisted that the claims that, for example, social inequality or competitiveness are built into human nature are very convenient for upholders of the capitalist status quo. As a result, said Marxists are often keen to deny such claims.

6.1.1 The caution of Richard Lewontin

One such self-professed Marxist is Richard Lewontin. His views are in some ways similar to those of developmental systems theorists. Like the developmental systems theorists, he rejects the idea that genes contain the information for building an organism; environment also plays an active role. In Lewontin's case, this view is connected to an overall view of science in general.

Ideology in science

One distinctive facet of Lewontin's view of science is that he thinks *ideology* plays a large role in shaping the beliefs of scientists, and the directions in which their research goes. He is a Marxist, and believes that scientific theories often have an ideological bias that reflects the interests of the dominant economic class. As an example, he considers IQ testing. According to Lewontin, the idea that there is something called "native" intelligence, that can be measured by standardized tests, is ideologically laden. IQ testing, he argues, promotes the idea that we have innate differences in mental abilities, which in turn helps to justify existing inequalities in society:

The problem for bourgeois society (and for socialist society as well) is to reconcile the ideology of equality with the manifest inequality of status, wealth, and power, a problem that did not exist in the bad old days of *Dei Gratia*. The solution to that problem has been to put a new gloss on the idea of equality, one that distinguishes artificial inequalities which characterised the *ancien régime* from the *natural* inequalities which mark the meritocratic society.   (Lewontin 2000b: 17)

He is not claiming that differences in mental ability do not exist; nor that genes play no part in bringing about those differences. He is claiming that the relationship between genetic and environmental factors in determining mental abilities is complex. It is not simply a case of: if you have genes *x*, you will be more intelligent than someone with genes *y*, no matter what. More generally, Lewontin thinks that the idea that genes play a unique role in shaping organisms is driven by political ideology. His book *The Doctrine of DNA* (1993) is subtitled *Biology as Ideology*.

Metaphors in science

A second distinctive feature of Lewontin's view of science is that he argues that science constantly employs metaphors. To describe light as a wave is a metaphor, as there is no medium for it to be a wave in. (It is not like waves in water or in air.) Such metaphors are extremely useful; indeed, science could not possibly dispense with metaphors. But metaphors can be misleading. They can lead us to have a distorted picture of reality. They can be useful in helping a new science to develop, but they can also outlive this usefulness.

Sometimes metaphors can shape our thinking without our realizing that they are doing so. An example in biology, he believes, is the concept of *development*. The idea of development implies that something that was, in some sense, *already there*, is made visible. Think of *developing* a photograph. The undeveloped film already contains the picture: the process of development just makes it visible. The picture that results is determined by what is on the undeveloped film. The idea that development makes visible what is already there can be seen in the seventeenth-century theory that sperm cells contained a "homunculus": a microscopic person. In the womb, it was thought, this microscopic person just got bigger. We now think this idea is clearly ridiculous, but, according to Lewontin the idea that genes are what shape an organism is essentially the same. What both ideas have in common is the idea that the shape of an organism is somehow already present right from the beginning.

From the arguments of Lehrmann and the developmental systems theorists, we can see why this view might be thought problematic. Lewontin's own view has a further twist that takes it beyond their views. According to Lewontin, the form of an organism is determined by genes *plus* environment *plus* chance factors.

Environment

Lewontin's argument for the active role of environment on development is a straightforward empirical one ("Gene and Organism" in Lewontin 2000a). Even if we acknowledge that how genes are expressed is affected by environment, we still might think that the relationship between the genotype and the outcome is a straightforward one. For example, let us say that Mary has genes "for" being tall, while Maria does not. We might think that, even though the exact heights of Mary and Maria will be influenced by the environment, they will be influenced in the same way. Environmental conditions *A*, we might expect, will make them both taller, while environmental conditions *B* will make them both smaller. Thus, if they are brought up in the same environment, Mary will be taller than Maria, no matter what that environment is.

In fact, actual studies on how environmental conditions affect size have been carried out on plants. Seven different *Achillea* plants were cloned, and the clones grown in three different environments. The result was not like the situation described above. The genotype that was the tallest when grown at a low altitude was also the tallest when grown at high altitude, but it was the smallest at a medium altitude. The one that was the tallest at a medium altitude was medium-sized at low altitude and small at high altitude, and so on.

In other words, we might expect things to be as in Figure 6.1. But what we actually find is shown in Figure 6.2. There is no straightforward correlation between genotype and height at all. (To put it in more technical terms, the *norm of reaction* is not a straight line, but a highly irregular curve.)

We can now see why Lewontin is sceptical about there being such a thing as native intelligence that can be measured by IQ tests. It was just *assumed* by the designers of IQ tests that IQ would be affected by environment in the way depicted by Figure 6.1. But this, according to Lewontin, is based on an unrealistic view of how environmental changes actually affect the expression of genes. In actual fact, there is still no agreement on whether there is a strong correlation between genes and IQ across different environments; empirical evidence has so far given us no conclusive answer.

Chance

Lewontin also suggests that there is a role for *chance* in how organisms develop. This is because micro-level differences between individual cells can have macro-level effects. He gives the example of the sensory bristles on *Drosophila*. The formation of each one of these bristles depends on just three cells, which have to be at the surface of the fly's body *at the right time* in the fly's growth. They have to be there before the body surface hardens. But this, in turn, depends on the three cells being formed by the division of one original cell *on time*. But the speed at which this cell division takes place depends on micro-differences in the chemical constitution of the original cell. If the original cell takes too

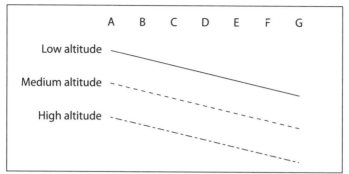

Figure 6.1 Correlating genes and height: what we might expect.

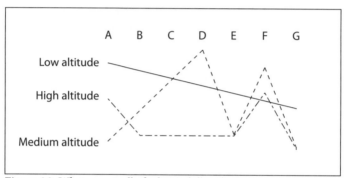

Figure 6.2 What we actually find in *Achillea* plants.

long to divide, no bristle is formed. These bristles are distributed differently on different flies. And this, Lewontin argues, is not due to any difference in *either* genes *or* environment. It is a result of random factors. It might seem that this is a trivial difference between individual fruit flies. But the bristles are not just for decoration; they are sensory organs.

Lewontin suggests that similar random factors may explain other differences between individuals of the same species. Perhaps, he suggests, when the brain is developing, neural connections are partly formed by random growth, and then either reinforced by external inputs or not. This is a kind of "natural selection" in the brain. The effect of the environment plays a part in whether some neural connection is reinforced or not. But the formation of the connection is itself random. This idea was proposed by Gerald Edelman (1987), who called it "neural Darwinism". Lewontin suggests that something like this could help account for differences in ability between people, such as musical ability:

> I am certain that even if I had studied the violin from the age of five, I could not play a Paganini caprice as Salvatore Accardo does, and

> Accardo no doubt has neural connections that I lack and has had them since an early age. But it is by no means clear that those anatomical differences between us are genetic. (Lewontin 2000a: 38)

What he is saying is that these differences may be *neither* genetic *nor* environmental.

So, according to Lewontin, the picture of the organism being somehow contained in its genes is shown to be misleading through the fact that: (i) there is no straightforward correlation between genotype and certain traits, such as height; and (ii) randomness plays a part in determining differences between individual organisms. The picture of an organism being somehow contained in its genes is, according to Lewontin, a metaphor. He is not denying that it has its usefulness, but it is misleading and obscures various possible lines of research. For example, regarding the suggestion about random factors influencing differences in people's abilities, he writes: "To relate the undoubted existence of random nerve connections to variation in specific characteristics like musical ability would require a major research program. But such a research program will only be carried out if the question is asked in the first place" (*Ibid.*). By stressing the role of chance in development, Lewontin goes one better than the developmental systems theorists.

6.2 But what is innateness?

However, a further question arises, which is the main subject of this chapter. When it is claimed that such-and-such a trait is innate, what exactly is being claimed? It will be noticed that in the previous section I used the terms "innate" and "part of human nature" as if they were interchangeable. Moreover, as I already indicated, other terms seem to be also substitutable, such as "inborn", "in our genes" and "part of our evolutionary heritage". Likewise, the contrast class seems to go under a number of different names: "acquired", "socially constructed", "learned" and so on. When we think about the dichotomy, a whole host of connotations spring to mind (Table 6.1). There are doubtless other items one could add to this list, which does not pretend to be exhaustive. (Indeed, at the end of this chapter I shall suggest that it *could not* be exhaustive.)

Looking at these lists of allegedly synonymous terms, we can make three observations:

- Despite the diversity of terms on each of the two lists, we have a sense that they all mean more or less the same thing. In other words, the terms seem to *pick out a single concept.*

Table 6.1 Connotations of the "innate–acquired" dichotomy

Innate	Acquired
Present at birth	Not present at birth
Product of genes	Product of environment
Bound to develop no matter what	Only develops in certain environments
Difficult or impossible to change	Relatively easy to change
Explained by biological evolution	Explained by culture/history
Biologically adaptive	Might or might not be adaptive
Typical of the species	Culturally specific

- This concept seems *perfectly familiar* to us, even to those of us who lack biological knowledge, let alone philosophical reflection on such knowledge. When we say that a trait is innate, is it not obvious what we mean?
- However, when we look a bit more closely at the two lists of allegedly synonymous terms, it becomes less obvious that they actually are synonymous. For example, the terms "product of genes" and "explained by biological evolution" will not seem interchangeable to anyone who finds developmental systems theory plausible. On that view, as we saw in Chapter 5, evolution has bequeathed to us a recurring life cycle, which includes, and is supported by, a host of interactions between genes, other parts of the organism, and various bits of the organism's physical surroundings. Moreover, developmental systems theorists hold that the expression "product of our genes" is meaningless anyway. But this rejection of the notion "product of our genes" is not confined to developmental systems theorists. As we also saw in Chapter 5, statements to the effect that we must attribute *all* traits to genes and environment together are commonplace in textbooks of evolutionary psychology and related disciplines.

Griffiths (2002) has argued that the concept of innateness belongs to *folk biology*. That is, it is a concept that appeals to us before we think rigorously about what scientific biology entails. In this respect, he says, it is like the idea that biological species are fixed for all time and have clearly defined boundaries. Someone who did not understand (or who did not accept) the theory of evolution might say "something is either a dog or not a dog", but the theory of evolution implies that that is false (see Chapters 8 and 9). Likewise, according to Griffiths, a proper understanding of how traits actually develop in an individual organism should lead us to abandon the concept of innateness.

The reason for this, he argues, is that as the term is usually employed, it carries a whole host of different connotations. When we decide that a trait is innate, we often infer various other things about the trait. We tend to infer, for example, that it is species-typical (i.e. in human beings, that it is invariant across different cultures), that it is difficult or impossible to change, that it is caused by the action of a gene, that it is a product of evolution (which in turn often leads to the further inference that it is an adaptation) and so on. But according to Griffiths these are distinct things, and there is no concept x such that, if a trait is x, then it follows necessarily that that trait is species-typical, difficult to change, caused by a gene, an adaptation and so on. He proposes that we should simply abandon the term "innate" altogether.

Most biologists and philosophers of biology agree that the word "innate", as it is used by the biologically uneducated, is very imprecise. This by itself, however, does not establish Griffiths's claim that we should abandon the word. In order to evaluate his claim, we should look at some attempts that have been made to formulate a more precise concept of innateness.

A number of people have attempted to capture the concept of innateness in a formulation. For the most part, especially in the past few years, they have generally admitted that no single formulation can capture all that seems to be entailed by the nebulous traditional (or "folk") concept. But some features that seem to be entailed by that nebulous concept are now generally admitted to be not worth capturing. Some do not seem to be worth capturing because they are incapable of being fulfilled by anything. For example, no trait can meaningfully be said to be determined *solely* by genes; no trait is unaffected by environment. Some do not seem to be worth capturing because they are of no real significance. For example, it is of no real significance whether or not a trait is manifest at birth. (The colour of your beard, if you are a man, is usually said to be innate, but you do not have a beard at birth.) But even leaving those aside, most recent formulators admit that their particular formulation requires some of the implications of the traditional concept of innateness to be jettisoned. I shall deal briefly with an account that is pretty clearly problematic, before going on to two more promising accounts.

6.3 The ordinary-language concept

The word "innate", as we use it in ordinary language, seems to carry a host of different connotations, which may not bode well for any attempt to give it a precise formulation. Nonetheless, we usually feel that the word means *something*, so perhaps the attempt is worthwhile. Perhaps the most notable attempt to capture all the connotations of the word comes from Stephen Stich, whose account of innateness can be found in the Introduction to his book *Innate Ideas* (1975), which gathers together classic and contemporary writings that make, or deny, claims about various things being innate. For example, it includes relevant excerpts from Plato's *Meno*, the debate about innate knowledge between Locke and Leibniz, and discussion of Chomsky's claim that certain features of human language are innate. He rightly notes that modern debates have usually been carried out without anyone giving a clear formulation of what the word "innate" is supposed to mean. This appears to be a symptom of the fact that I alluded to above, that we commonly take it to be *obvious* what it means, and hence do not feel any need to make it explicit. Stich, however, sets out to formulate it in words.

One of the first things he notices is that although we think of "innate" as entailing that something is present at birth, we in fact apply it to things that are not *literally* present at birth. An obvious example is the colour of your beard, as mentioned above. Stich also instances diseases that we call innate, yet that do not show themselves at birth. For example, Huntington's chorea does not usually show itself until the age of 35 or over, but is called innate for all that. The key phrase here, however, is *show itself*. We might say that something is present although it cannot be seen, as a cancerous tumour may be present even though it produces no visible symptoms. The disease is *present* in that the tumour is present, but it is not yet *manifest*. Similarly, we might say that a disease that is caused by an abnormal gene is present, in that the gene is present, but it is not yet manifest. Both the tumour and the gene are entities that are present in the perfectly ordinary sense of present; they can, for example, be seen if you have the right instruments. So, we can hold on to the notion of "innate" as "present at birth", by distinguishing between this and being manifest at birth.

But something is slightly awry here. What is present at birth is a *gene*. But we find it natural to say, at least sometimes, that the *outcome* is innate, not just the gene itself. Once again, we say of the colour of your beard that it is innate. We need to specify just what it is about having the gene that makes it the case that certain outcomes of having that gene are innate. As a first pass, Stich suggests that something is innate if the state of the person (or animal, or whatever) at the beginning of life is such that, if they were at a certain stage of life they would have that trait. And here, he means that the trait itself would be present in the perfectly ordinary sense of present, not just that some gene is present. This works, apparently, for the beard example, as it is indeed the case that if the person were at the right stage in life he would have a beard.

One consequence of this is that what one then counts as innate will depend on what one counts as the beginning of life. If one takes it to begin at conception, and if one also takes having the gene for a disease to be the sole and sufficient ground for saying that one has it innately, then Stich's proposal seems to work. However, having a gene *never* by itself makes it the case that if someone is at the right stage of life he or she will definitely have a certain trait. The delicacy of the developmental process when the foetus is in the womb clearly rules that out. Moreover, recall Lewontin's suggestion that differences in, for example, musical ability may be due to random connections being set up in the brain during its early development, which profoundly shape later development and are to all intents and purposes irreversible. We might plausibly, as Lewontin himself suggests, call such differences innate, in that the conditions that create them are present extremely early in life (probably before birth), they are not due to the environment and they are difficult or impossible to change. Be that as it may, Stich opts to call the beginning of life "sometime before birth", and accordingly is prepared to accept characteristics that are due to events in the womb as innate.

However as the definition stands it is too strong. Suppose someone has the gene for some condition, but before she reaches the right age a cure is developed, so she never gets the symptoms. We would still say, presumably, that she has the disease innately, at least up to the time the cure is given to her. As a way of dealing with this, Stich adds the proviso "in normal conditions". So his final formulation is: "A person has a disease innately at time *t* if, and only if, from the beginning of life to *t* it has been true of him that if he is or were of the appropriate age (or at the appropriate stage of life) the he has or in the normal course of events would have the disease's symptoms" (Stich 1975: 6).

A word of explanation is in order about the proviso "from the beginning of life to *t*". Stich wishes to hold that, if the cure has not yet been given, the person has the disease innately, even if she does not yet have the symptoms. But, apparently, he also wishes to hold that, after the cure has been given she does not have the disease innately. Hence the proviso. Be that as it may, we might wonder about the "normally" addition. The term "normally" is clearly vague. First, it does not specify *how* abnormal the conditions have to be in which the symptoms will not arise for the disease to be counted as innate. In practice, people with a gene "for" a disease are often told that they have a *high likelihood* of actually developing it. This quantitative vagueness, however, is not a fatal problem. As you will see in Chapters 8–10, we are used to dealing in biology with concepts that have fuzzy boundaries. Note that in the list of connotations of "innate", I put "difficult or impossible to change", thus allowing that innateness could be a matter of degree. As we shall see, other accounts of innateness make it a matter of degree as well.

The problems do not stop there, however. One thing that is left unaddressed in Stich's account is *why* a trait develops in normal conditions. Suppose "normal conditions" includes a particular nutrient that is regularly, reliably available. This nutrient, let us say, needs to be in a creature's diet for it to become a certain colour. The distinctive bright pink colour of flamingos is due to a diet that is high in alpha- and beta-carotene. By contrast, imagine a different creature that turns out to be the same colour on any of a wide range of different diets. Even if carotene was *always* present in flamingos' diet, we might want to say that there is a difference here between the reason why flamingos are usually pink, and the reason some other creature is usually the colour it is. We might not say that flamingos are *innately* pink. Or imagine that it becomes a normal feature of human beings that they eat too many fatty foods and do not exercise, so that obesity becomes extremely widespread. We would, again, want to distinguish between obesity and the fact of having 32 teeth, and say that the latter is innate and the former is not. The "normally" proviso does not distinguish between cases where a trait reliably develops just because the environmental resources used to develop it are invariable and ones where they reliably develop for some other reason. Yet we presumably would not call the first kind innate. "Innateness" seems to carry with it the connotation of being (relatively) insensitive to variations in the environment. And that has to mean more than just "insensitive to the variations in the environment that do not affect it", for by that definition every trait would be innate. In §6.4 we shall see an attempt to give more substance to this notion.

Stich's account is commendable for trying to capture our common-sense understanding of the word "innate". But his failure to do so may be a symptom of the futility of that enterprise.

6.4 Canalization

André Ariew (1999) makes use of the concept of "canalization", which was developed by C. H. Waddington in the mid-twentieth century (see Waddington 1975: chs 11–13). Waddington was led to develop the concept by experiments on early embryos. He showed that when ectoderm tissue is at the stage at which it can become neural tissue, it can be induced to do so by a large variety of different compounds, both natural ones and artificial ones (see Gilbert 2003: 391ff.). The implication is that the development of these plates does not depend on a narrowly specified set of environmental circumstances (i.e. the presence of one particular compound); so the trait can be relied on to develop in a wide variety of different environmental conditions.

This state of affairs is in contrast to one in which the trait has to rely on *one particular* set of resources to develop. Some traits are environmentally sensitive in this latter way; if they nevertheless usually develop in the organism's natural environment, it is because: (i) that environment is very stable with regard to the resources needed; or (ii) the organism moves around to track the resources; or (iii) the organism (or the population to which it belongs) manipulates the environment to ensure the resources are present and so on. (I do not pretend this list is exhaustive, but you should get the idea.) Thus we have a clear contrast between traits that reliably develop in a wide variety of *relevantly* different environmental circumstances, and ones that do not. A trait that reliably develops in a wide variety of different environments *in virtue of being able to employ a wide variety of different resources in its development* is said to be *canalized*.

Two consequences of this formulation should be noted. First, canalization is a matter of degree. No trait is canalized *simpliciter*, since there will always be some environmental conditions in which it will not develop (even if the organism develops normally up to the point where the trait "should" appear), but some traits are more canalized than others. Secondly, even this canalization-as-a-matter-of-degree is only relative to a range of environments. Nonetheless, that may be a broader or a narrower range.

Unlike a crude conception of innateness that sees it as entailing insensitivity to all and any environmental vicissitudes, the concept of canalization captures a feature that some biological traits genuinely have. Further, unlike a vacuous conception of innateness as *somehow* influenced by genes, or *somehow* a product of evolution, canalization captures a feature that some biological traits genuinely lack. In other words, even though it is only a matter of degree, and is only relative to a range of environments, the difference between being canalized and not being canalized is a *genuine* difference. It cannot be said that all biological traits are equally canalized. (Contrast this state of affairs with John Tooby and Leda Cosmides's claim that I quoted in Chapter 5: "*everything* … is totally and to the same extent genetically and environmentally codetermined".)

As already pointed out, any concept of innateness that implies that innate traits will *invariably* develop is far too strong. No trait could be insensitive to *any* environmental influence. It is nonetheless reasonable to expect traits to vary (relative to a given range of environments) in the *degree* to which they are sensitive to environmental influence, which is exactly what canalization captures. So, if we wish to defend a concept of innateness that is a matter of degree, and we wish to capture the distinction "sensitive-to-environmental-variation *versus* insensitive-to-environmental-variation", then canalization does the job. As I mentioned earlier, some people see innateness as intimately tied to the further feature of having an adaptive explanation. Canalization does not *entail* this, but the fact of a trait being canalized is likely to be, in many cases, a product of adaptation. If a trait is canalized, a likely explanation is that it was important to the organism and needed to be reliably developed in a variety of different environmental circumstances. (However, it is *not* the case that an adaptive trait is in all circumstances more likely to be canalized.) Moreover, empirical evidence supports this conclusion, as experiments – again carried out by Waddington – have shown that artificial selection for traits that were initially not canalized eventually led to those traits becoming canalized.

Ariew's account captures (or at least tries to capture) developmental fixity at the expense of species typicality and/or having an adaptive history. In defence of this, it can be argued that the traditional concept of innateness *seems* to apply as much to non-species-typical traits as to species-typical ones. We say that eye colour, or genetically inherited susceptibilities to disease, are innate. Moreover, the traditional concept seems to apply as much to non-adaptive traits as to adaptive ones. Once again, we call genetically inherited susceptibilities to disease innate.

6.5 Generative entrenchment

Wimsatt's (1999) view is in many ways similar to that of the developmental systems theorists. Like them, he does not believe that there is any meaningful distinction to be drawn between "replicators" (such as genes) and "vehicles" (such as everything else about an organism). Nonetheless, he thinks that there is something worth preserving in the traditional concept of innateness.

He proposes to capture this using the concept of *generative entrenchment* (see §5.3.1). To recap, features that are hard to change without having to redesign all the other things that are "built on top of" them are said to be generatively entrenched. Here, with a little explanatory preamble, is Wimsatt's definition of the term:

> [U]ltimately, in evolving systems, it *is* the generative role of elements that causes resistance of their changing. The anchoring against change of structures of widely used elements requires few assumptions:
> (1) structures that are generated over time so they have a developmental history (*generativity*), and
> (2) some elements that have larger or more pervasive effects than others in that production.
> Different elements in the structures have downstream effects of different magnitudes. The *generative entrenchment* of an element is the magnitude of those effects in that generation or life cycle. (1999: 148)

Note that generative entrenchment, like canalization, is a matter of degree. Our particular kind of gastrulation is more deeply entrenched than the tetrapod skeletal structure, and is common to all deuterostomes, including fish, sea squirts and starfish as well as tetrapods. At the other end of the scale, the universal grammar that is common to all (or nearly all) human beings, according to Chomsky, is more deeply entrenched than the particular language you speak.

Wimsatt does not say that innateness *is* generative entrenchment. In fact, like Griffiths, he writes that "the old innate–acquired distinction should be retired", but immediately follows this with "but its conceptual niche is not dispensable and is filled fruitfully by the new concept" (*ibid.*: 138) Later he writes: "[generative entrenchment] can be used to give a powerful analysis of phenomena that the innate-acquired distinction has been invoked to explain throughout its range in ethology, cognitive development, linguistics, philosophy, and elsewhere" (*ibid.*: 152). He is keen to show that most of the connotations of the traditional concept of innateness are captured by generative entrenchment. He goes to the trouble of listing twenty-eight such connotations, arguing that generative entrenchment captures most of them.

Generative entrenchment is certainly an important concept in thinking about evolution. Moreover, it does seem to fit many of the cases where we would normally use the word "innate". According to Chomsky, universal grammar is innate, just as having thirty-two teeth is, which explains why it is universal. By contrast, even if everyone in the world spoke English, we would still not say it was innate. So, if Chomsky is right, universal grammar belongs on the same side as having thrity-two teeth, and not on the same side as speaking English. Wimsatt's account accommodates this distinction: if Chomsky is right, then even if English became universal, it would still be easier to lose English without losing the universal grammar than it would be to lose the universal grammar without losing English. And this is because the universal grammar stands in relation to specific language as foundation to superstructure. No matter how widespread any language became, we would not say it was generatively entrenched in the way that universal grammar is.

But generative entrenchment does not line up with the traditional concept of innateness perfectly. On reason is related to the fact I mentioned earlier: that we apply the word "innate" to some traits that appear late in development. Wimsatt is aware of this objection, and argues that generative entrenchment does not require that something appear early in development. The behaviour of parents towards their young is influenced by "imprinting", yet imprinting is something that happens to the parent when it is an adult. But that does not deal with the beard example I mentioned earlier. We call some traits "innate" even if they have no effects on future development, as, for example, the colour of your beard does not seem to.

Secondly, species typicality does not seem to be necessary for innateness as the word is usually used. The concept of generative entrenchment does not *require* it either, but there is an explanatory–predictive link between generative entrenchment and species typicality. That is, if a trait is generatively entrenched and species-typical, that is no accident. We would expect traits that are generatively entrenched to be species-typical *at least*: indeed, those that are very highly generatively entrenched we would expect to be typical of larger groups than species, as we saw. Wimsatt is aware of this link, and in fact considers it an advantage of his view. He attaches great importance to satisfying the criterion of species-typicality:

> It … becomes all too easy to speak of innate intraspecific differences in all kinds of traits. But we've been here before, and it is dangerous and easily misused territory. Chomsky's scientific tastes for species-

> universals correspond with a good place to draw the distinction: keep
> innate differences at the species-level or above. (*Ibid.*: 165)

The phrase "dangerous and easily misused territory" is presumably meant to make us think of "racial science" and other evil consequences that one could draw if one thought that non-species-typical traits could be innate. But the key word here is "could". Surely only if one thinks that *some* intraspecific traits, such as skin colour, are linked to *some* others, such as intelligence or moral character, is one going down that path. And one can deny that while believing that there are some innate intra-specific differences. There seems to be no escaping from the fact that eye colour, hair colour and various hereditary diseases are as clear examples of innate traits as anything else. The first two do not seem to be generatively entrenched, as they do not seem to require taking other things apart to change them. You can dye your hair without too many knock-on effects. The hereditary diseases would be better described as disruptions of things that are generatively entrenched, rather than generatively entrenched themselves.

6.6 A deflationary approach

In a very recent sally on the problem, James Maclaurin proposes to cut away all the "ifs", "ors" and subordinate clauses, by going straightforwardly for a single simple criterion: "[M]y proposal is that we reinvent 'innateness' as an umbrella term picking out all the traits that are, by whatever mechanism, built-in to populations" (2002: 109). The key phrases here are "built in" and "by whatever mechanism". Like developmental systems theorists and evolutionary psychologists, he rejects the idea of a trait being fully determined by a gene. Indeed, he cites Lewontin's evidence that the way genes and environment interact to produce outcomes often follows highly complex, non-linear patterns. However, Maclaurin does not want to go as far as to say that there is never any difference in the relative contributions of genes and environment. He thinks it does make sense to say that some traits are "more genetic" than others. But that, he thinks, is of no real significance. Traits can be built in to a population for many different reasons: they may be built in largely due to the homeostatic effect of a gene, or they may be built in because the relevant developmental resources in the environment are highly stable, or because the trait is needed if the organism is to live at all. In any of these circumstances, he holds, the trait can be considered built in and therefore innate.

"Built in" means, and therefore "innate" means, nothing more than "typical of a population". There is, admittedly, a twist to this. He considers that a disease such as Huntington's chorea counts as built in, even though only a tiny percentage of people have it. This is because Huntington's chorea is a regularly recurring phenomenon in the population, and develops not *just* because of a gene, but because of that gene *in conjunction with* stably recurring features of the human species's world. The very same things that ensure that most of us develop normally are also implicated in a person with the gene developing the disease.

Often the relevant population is the species (indeed, some taxonomists think that "evolutionarily significant group" is the definition of "species"), and Maclaurin's examples suggest that he thinks that, as far as human beings go, it usually is. In terms of my list of the connotations of the traditional concept, then, Maclaurin's account picks out species typicality at the expense of linkage to a gene and difficulty of change. This is not as indefensible as it probably first sounds. In its defence it might be argued that often the significance of innateness lies in identifying traits as products of evolution, rather than as having a particular type of developmental history in individual organisms. This suggestion gains some plausibility from the fact that if a trait exists at high frequencies in different cultures – or, more generally, in different groups of a species in varied environments – that is a good *prima facie* reason to believe it is innate.

However, *prima facie* evidence for something is not that thing itself. It may be that, as most linguists believe, there are innate features of language, and these clearly fit the criterion of species typicality (not to mention hearts, kidneys, eyes, ears, etc.). But a trait could be species-typical without being what we would normally call innate. Suppose that the English language were to become the dominant language of the world, with all others reduced to tiny pockets of speakers, as, for example, the Irish language is today. We would still not say that the English language was innate. Surprisingly, Maclaurin is prepared to live with this consequence of his account. In fact, on his view a trait does not *have* to be typical of the species to be innate. So English, as well as a person's religion, can count as innate even today:

> Breaking my thumb at the age of eight is not innate as there is no mechanism in my population that regularly produces such outcomes in my lineage. However, there are mechanisms that maintain all sorts of behavioural characteristics in human populations. So on my story, highly stereotypical, inherited cultural phenomena (such as religious beliefs) could turn out to be innate. (2002: 126)

But even if we grant him this, a further problem is that Maclaurin's account makes typicality *tout court* the replacement concept for innateness. As he says, something can be typical for many different reasons. But innateness is usually treated as if it were an *explanation* of why something is there. If we asked "Why does so-and-so have X?", we might get the answer "Because it is innate". But if the answer was "Because it is typical of so-and-so's population", we would perhaps feel a bit short-changed. Surely we would want to know more than that most other things like so-and-so have X as well. (If this were a strict law that might be different, but "typical" is not a concept that is all-inclusive enough for that.) The answer "Because of its genes" at least aspires to being an explanation, even if it is problematic. For all the answer "Because it is typical" tells us, it could be typical in the way that speaking English would be in my imaginary scenario, or typical in the way that having thirty-two teeth is. But these have utterly different kinds of explanation. Maclaurin is, of course, fully aware that that they do, but the word "innate" is usually intended to *mark* this difference in explanation. On his account, it no longer does so.

Ultimately, Maclaurin's account looks like an instance of what philosophers call *deflationism*: that is, taken a concept that is generally held to be laden with important implications, and redefining it in terms that lack most or all of those implications. Maclaurin fully understands that the traditional concept is very unsatisfactory, and that the cascade of implications that people often attempt to derive from it should be avoided. But, given that his account is *very* deflationist, it is not clear why he wants to retain the word "innate" at all.

6.7 Conclusion

The accounts of Ariew, Wimsatt and even Maclaurin all successfully capture some of the connotations of innateness. Moreover, they all capture features that a trait can genuinely have, unlike, say, being completely determined by a gene, or being completely unaffected by the environment. There really is such a thing as canalization, there really is such a thing as generative entrenchment, and there really is such a thing as species or population typicality. And these features are all of genuine biological interest, unlike, for example, whether or not a trait is manifest at birth. Why not, then, take one of them as the real meaning of the word "innateness"? It is, after all, not unheard of for science to stipulate a precise definition of a term, even at the cost of excluding some things that seem to be instances of what the term refers to. At the time of writing this, the International Astronomical Union and the Division of Planetary Sciences of the American Astronomical Society have just voted on the criteria for deciding what constitutes a planet, in order to decide whether the recently discovered object UB313 is one. An outcome of their deliberations is that Pluto ends up not being a planet. Absurd though this particular situation may be, the need to *sometimes* make decisions like this is a product of a perfectly reasonable desire for precision in science.

But if we decided to take one of the rival accounts as the true definition of innateness, how would we decide which one to pick? While they all capture some of the connotations, none of them capture all of them. Perhaps we could put together a list of all the connotations, and pick whichever account captures most of them. There are many connotations; Wimsatt lists twenty-eight, claiming that generative entrenchment captures most of them. The fact that there are so many of them might even be thought to be a good omen, as it reduces the chances of a tie. Unfortunately, not only is the list of connotations long, but it is also *nebulous*. It is not clear exactly how many items go on the list, partly because it is not clear whether innateness does connote some of them, and partly because it is not always clear if two of them are really distinct. Among the items on Wimsatt's list is "universal within species", which I have already argued is not really a connotation of the word "innate". He also includes "resistance to evolutionary change" and "evolutionarily conservative", which looks like the same thing under different names. So the problem cannot be solved by counting connotations.

As an alternative, why not simply accept more than one definition of innateness? We could call them "innateness A" and "innateness B". Strange though this may sound, in Chapter 9 we shall see that some people recommend doing precisely this with the word "species". But in the case of innateness that would be a bad idea. The reason for this is that the folk concept seems to go very deep. In particular, when we hear that something is *innate*, we tend to think that that implies that it is in some sense present at birth, that it is in some sense a product of genes (note the nebulousness!), that it is bound to develop no matter what and so on. Moreover, when we hear that something is one of those things, we tend to infer that it is also all the others. Evidence of this can be seen in the quickness with which people react to evolutionary psychologists' claims about certain traits being part of our evolutionary heritage. Admittedly, it needs to be spelled out in more detail what exactly the evolutionary psychologists themselves take it to imply, and to what extent this matches traditional claims about things being "innate". I shall defer this task until Chapter 13. But, as I indicated in §6.1, there is a strong tendency for people to react by inferring that it means those traits are genetically determined, difficult or impossible to change and so on.

This in turn often leads to moral and ideological outrage, which can hardly be said to be the upshot if someone claims that the salamanders of the western USA are a single species. For this reason, it should be kept clear that the different concepts associated in the popular mind with innateness are just that, *different* concepts, each of which does not necessarily entail the others. Practitioners in various disciplines would be well-advised to avoid using the word "innate" at all because it seems to lead to a cascade of illicit inferences. This also gives us a further reason why we should not try to define innateness in terms of one of the other concepts. Better to just say canalized when you mean canalized, species typical when you mean species typical and so on. Hence, I agree with Griffiths when he writes:

Substituting what you actually mean whenever you feel tempted to use the word "innate" is an excellent way to resist this slippage of meaning. If a trait is found in all healthy individuals or is pancultural, then say so. If it is developmentally canalized with respect to some set of inputs or is generatively entrenched, then say that it is. If the best explanation of a certain trait difference in a certain population is genetic, then call this a genetic difference. If you mean that the trait is present early in development, what could be simpler than to say so?

(Griffiths 2002: 82)

As we saw, evolutionary psychologists are very careful not to use the word "innate". But their opponents still persist in imputing views to them that only make sense if you accept the full gamut of connotations that the traditional concept of innateness carries. Evolutionary psychologists say that they see no meaningful distinction between "innate" and "acquired", and sometimes set out to explain what it is they are claiming about the evolved cognitive architecture (more on this in Chapter 13). Perhaps they should do both these things more loudly, more clearly and more often.

7. Function: "what it is for" versus "what it does"

It was perceived very early in the history of biology that living things have parts that are *for* things. Aristotle (arguably the founder of biology) fully appreciated it: "In dealing with respiration we must show that it takes place for such or such a final object; and we must also show that this and that part of the process is necessitated by this and that other stage of it" (*On the Parts of Animals*: Bk I, pt 1).

The English word "teleology" is derived from the Greek word "telos", which may be translated as "purpose". What does this mean when applied to living things? When we think about a spider building a web, or a squirrel gathering nuts, we cannot help but see purposefulness in these activities. The spider builds a web *in order to* catch flies; the squirrel gathers nuts *in order to* secure enough food for the winter. By contrast, we would not say that the acid eats away at the metal *in order to* bring about some end; the process merely has an end result, and that is all. When it comes to the activities of animals and plants, it seems much more natural to say that they are carried out in order to bring about some end. Why does this idea have such obvious appeal?

Analogies between human beings and other living things
One reason for the appeal may be because of apparent analogies between other living things and ourselves. We know that human beings' activities are (at least very often) *purposeful*. Often it is pretty obvious why people are doing what they are doing; if someone is putting on a raincoat, she is presumably doing so in order to avoid getting wet. But when we look at the activities of animals (leaving plants aside for the moment), they seem to be doing things that we also do: they eat, sleep, fight, run away from danger and so on. However reluctant we might be to call any of these activities, when carried out by animals, *intentional*, we cannot avoid seeing *some* similarity to our own.

It may be thought that our seeing of this alleged similarity is nothing but a habit, done unthinkingly and due perhaps to sentimentality or to picking up on superficial similarities. Attributing properties to animals that really should only be attributed to human beings is called *anthropomorphizing*. Clearly, if I said that my dog enjoyed Mozart or understood chess, most people would say I was anthropomorphizing. Some philosophers and zoologists would go further, and say that I am anthropomorphizing if I say that my dog is sad, or angry, or believes I will soon be home. There is wide disagreement over to what extent, and in what ways, human beings are unique.[1] But the instinctive gut feeling that we have that animals – even quite simple ones – are somehow "like us" seems hard to gainsay.

Perhaps the appeal of this justification for ascribing purposes to animals can be strengthened if we accept that ascribing a purpose even to a person is *not* the same as ascribing an intention to that person. Not all the purposes of human activities need be intentions of the human beings carrying out those activities. Suppose we were to ask: why is so-and-so eating? One answer might be: because he is hungry, and satisfying his hunger may be his intention when he eats. But an alternative answer might be: in order to stay alive. When we eat, we do not normally consciously think about the necessity of eating for staying alive, but it is still an answer that might be given. Of course, we know that we need to eat to stay alive, and we would generally prefer to stay alive than to die. But there is a distinction between a reason that one *would* endorse for doing something, and the reason that one actually has for doing something. If one had not eaten for, say, twenty-four hours, one's actual reason for eating would most probably be hunger rather than the desire to stay alive. In any event, we often do not have reasons, in the sense of thoughts of the form "if I do this, this will follow" for eating: we just eat. But that does not prevent it being the case that there are reasons, in the sense of purposes, for doing so. And while we might disagree over whether it is anthropomorphic to say that my dog is sad or has beliefs, it hardly seems anthropomorphic to say that it eats. Eating is clearly an activity with a purpose, even if it does not involve any intentions.

This enables us to see animals as having purposeful activities. We only have to take one small step more to enable us to take plants out of the brackets in which I placed them a little earlier. This is because many of the processes that go on inside plants – photosynthesis, absorption of water and so on – also serve the end that we said was the purpose of eating: keeping the organism alive. We do not have to say that the organism is *doing* anything in order to say that processes going on in it are purposeful. Indeed, many of the processes going on in our own bodies, as well as in those of other animals, can be seen in this way: for example, heartbeat, circulation of blood, digestion and so on.

What we are imputing is the idea that there is such a thing as "what is good for" the organism. We are familiar – from the world of advertising and popular media if not elsewhere – with the thought that walking four miles a day, or eating five portions of fruit or vegetables or whatever is "good for you". Similarly, I might say that a certain type of compost is good for the flowers in my garden, *in the same sense* of "good for": the sense, that is, of "conducive to health". The notion of health seems to go beyond mere survival, since a person might after all live to be 100 and be unhealthy for all that time. The word "healthy" calls to mind images of robust growth, strength, vigour and so on. One *might* want to reduce these notions to ability to withstand hostile environmental conditions, resistance to disease and so on and thus to increased chances of survival, but one might not. In any event, many of the processes that go on within animals and plants seem to be conducive to greater health, rather than just survival. Our immune system, for example, protects us against diseases that are not life threatening as well as against ones that are. Once one separates purpose from intention, it seems just as acceptable to say that such activities are *for the purpose of* promoting the health of the organism as are those of a conscientious gardener.

Analogies between machines and living things
A second possible reason why we see purposefulness in biological processes also stems from an analogy. Not only is it appealing to see other living things as like us, but it is also appealing – or at least it has been for many people for many centuries – to see them as being like objects made by us. A celebrated example of the use of this analogy can be found in Paley's *Natural Theology*:

> I know no better method of introducing so large a subject, than that of comparing a single thing with a single thing; an eye, for example, with a telescope. As far as the examination of the instrument goes, there is precisely the same proof that the eye was made for vision, as there is that the telescope was made for assisting it. They are made upon the same principles; both being adjusted to the laws by which the transmission and refraction of rays of light are regulated. I speak not of the origin of the laws themselves; but such laws being fixed, the construction, in both cases, is adapted to them. For instance; these laws require, in order to produce the same effect, that the rays of light, in passing from water into the eye, should be refracted by a more convex surface, than when it passes out of air into the eye. Accordingly we find that the eye of a fish, in that part of it called the crystalline lens, is much rounder than the eye of terrestrial animals. What plainer manifestation of design can there be than this difference? What could a mathematical-instrument-maker have done more, to show his knowledge of his principle, his application of that knowledge, his suiting of his means to his end; I will not say to display the compass or excellence of his skill and art, for in these all

comparison is indecorous, but to testify counsel, choice, consideration, purpose? (Paley 1809: 18–19)

Elsewhere in the same book, Paley compares bodily organs to a watch. We can also see the organism–machine comparison if we think about the attitude taken by a surgeon towards the human body. A certain amount of professional distance is appropriate, and a surgeon tinkering with, say, a person's heart is in many ways like a mechanic tinkering with a person's car. In both cases the expert knows what is where, what each part is supposed to do and (hopefully) what to do about something that is not working properly. Dawkins acknowledges that both the title and the content of his book *The Blind Watchmaker* are inspired by Paley. This parallel between biological traits and human artefacts has been developed by Dennett (1995).

One way we might describe this is to say that biological systems are highly *ordered*. It has often been pointed out that the mechanisms for transcribing genes are orders of magnitude more exact than any human copying down a text. The heart possesses mechanisms for regulating the pumping of blood in response to needs for more at one time, less at another. Mammals possess mechanisms for keeping body temperature constant despite changes in the ambient temperature. All these, and many more, biological mechanisms are very good at keeping *a predictable pattern of events* going; they preserve order.

The physical sciences proceed on the assumption that the universe works in an ordered way, and they have a very impressive track record in uncovering such order at many levels: the constancy of the speed of light; the recurring patterns in the periodic table of the elements; and so on. However, the order found in biological systems seems to be *over and above* that uncovered by the physical sciences. A fully adequate physics might tell us all about the order at the most fundamental level of matter, and might moreover lead us to expect order at higher levels, such as those of chemistry, as well. But it would not, as far as we can tell, lead us to expect the order we find in biological systems. It would not, as far as we know, violate any law of physics if biological order had never come into existence. Indeed, as far as we know so far, it has only come into existence on this one planet. This is different, of course, from saying that physics *cannot explain* how the ordered biological systems work. Similarly, it would not violate any law of physics if watches and computers had never come into existence, even though physics can explain how watches and computers work. The purposefulness and the order of biological systems seem to be genuine features of the world, and not just our own projections onto it.

Philosophers of biology often ask: what does it mean to say that such-and-such is the function of something? There seem to be three things that an answer to this question should try to capture:

- The word "function" means the same thing when we talk of the function of a part of an organism as when we talk of the function of a part of a human artefact.
- Its meaning has something to do with purpose, teleology, being *for* something.
- The functions of parts of organisms are explained by natural selection, whereas the functions of parts of human artefacts are explained by conscious design.

One of the most influential attempts to answer this question was made by Larry Wright ([1973] 1994). Wright's account attempts to preserve the idea of function as "what it is for", and the crucial link between biological function and adaptation. Ruth Millikan's account (1984; 1989) also attempts to capture the "what it is for" aspect. However, she abandons the attempt to find a concept of function that is common to biological traits and human-made artefacts. The opposite is the case with the account of Robert Cummins ([1975] 1994). He sets out to preserve the commonality between the two different cases of function, but argues that no viable account of function in terms of "what it is for" can be given. Accordingly, he argues that the appropriate use of the notion of function in biology is in terms of "what it does".

7.1 What it is for

The term "function" seems to carry the connotation of "purpose", or something being "for" something. We say that the heart's function is to pump blood, and we also say that the heart's *purpose* is to pump blood, or that the heart is *for* pumping blood. The three statements seem to be saying the same thing. In the 1960s a number of philosophical accounts of function generally attempted to capture this (e.g. Canfield 1964; Beckner 1969). However, by far the most influential account of function along these lines has been that of Wright ([1973] 1994).

Wright set out to give an account of function that would respect a number of (what he took to be) fairly obvious facts about functions.

(i) *The concept of function that we apply to parts of living things is the* same *concept that we apply to human-made artefacts.* Just as we say that the function of the heart is to pump blood, we also say that the function of the second hand of a watch is to make seconds easier to read. As I pointed out above, the analogy between the parts of organisms and those of man-made artefacts is a perennially appealing one. It seemed obvious to Paley and others, who used it to argue that the exquisite design of, for example, the eye, is evidence of the work of an intelligent designer. Even among those who think that natural selection rather than God produced the parts of organisms, the analogy between them and artefacts is often taken to be obvious. For example, Dennett quite happily talks of the "design work" of natural selection, and compares biology to engineering (Dennett 1995). The fact that the parts of organisms have functions, then, may be considered a *datum*, for which God (i.e. a conscious designer) and evolution are often taken to be rival explanations. That, together with the fact that we seem to mean the same thing by "function" when we speak of the function of a watch and the function of an eye, suggests that an account of function should be neutral between evolved entities and consciously designed ones. "Functional ascriptions of either sort have a profoundly similar ring" (Wright [1973] 1994: 30).

(ii) *Something can have function F, even if it is never actually used for F.* The second hand of a watch may never be used during the watch's lifetime to make seconds easier to read, but that is still its function. Likewise, we might have a reflex that causes us to jump out of the way when we see a snake. Even if a particular person never encountered a snake in her lifetime, it would still seem natural to say that the function of the reflex in her is to avoid snakes.

(iii) *The function of something is distinct from a mere accidental effect it has, even if that accidental effect is useful in some way.* The second hand may have the effect of removing bits of dirt, but that is not its function, no matter how useful it is. A further example that Wright gives is the buckle on a cowboy's belt that saves his life by stopping a bullet. We would still not say that that was the buckle's function. We might add, if the belt-maker deliberately put on an extra-large buckle specifically to stop bullets, then stopping bullets *would* be the buckle's function, or at least one of its functions. We want an account that will respect the difference between these two cases. For a biological example, the "lub-dub" sound that the heart makes is useful to us, because doctors can listen to it. But we do not say that making that sound is its function.

(iv) *A thing can have a function even if the system of which it is a part would work perfectly well without it.* If the second hand of a watch was not there, the watch would still work. Contrast this with one of the internal cogs of a watch, without which the watch would not work. Despite this difference, in both cases we say that it has a function. Likewise, eyebrows have the function of keeping sweat out of the eye but, unlike our hearts, our eyebrows do not keep us alive. In any event, we cannot say that something has a function only if it is necessary for the system of which it is a part to work; strictly speaking, that is never true of anything. If I did not have a heart, it is possible that something else could do the same job: an artificial heart, for example.

(v) *A thing still has the function it has even if it is incapable of performing it because something has gone wrong.* A malfunctioning heart still has the function of pumping blood, even if it is incapable of pumping blood. Indeed, the very idea that something is *malfunctioning* presupposes that it has a function. The second hand on a watch may be defective from the day the watch was made, and so never do what it is supposed to do. Even so, we would still say that enabling us to tell seconds is the second hand's function.

(vi) *The function something has plays a role in explaining why it is there.* In the case of a designed artefact, it was put there with the intention of performing that function. In the case of an evolved trait, it was selected because it performed that function. The "because" here, Wright notes, is *aetiological* because it indicates why a thing *got to be there*. In that respect, it is like the "because" in "it exploded because it got too hot" (*ibid.*: 39), not like the "because" in "it is hot because it is red" (*ibid.*). Of course, the "because" in "it exploded because it got too hot" is different from a functional explanation in other ways. The distinction between functional explanations and straightforward causal explanations is, Wright says "a distinction *among* etiologies; it is not a contrast between etiologies and something else" (*ibid.*: 41).

Wright examines a number of earlier accounts of function, and finds them failing to meet one or other of these criteria. For example, John Canfield defines function as "A function of *I* (in *S*) is to do *C means I* does *C* and that *C* is done is useful to *S*" (Canfield 1964, quoted by Wright [1973] 1994: 31). This falls foul of requirement number (iii); on this definition, stopping bullets is a function of a cowboy's buckle, and a function of your heart is that your doctor can listen to it. Moreover, it falls foul of requirement number (ii); if a second hand is never used to tell seconds, then on this definition that is not its function. Morton Beckner (1969)[2] gives a long and elaborate definition, which I will not give in full here. One of its components is "*F* [the function of *P*] is or contributes to an activity *A* of the whole system *S* [of which *P* is a part]" (paraphrased by Wright [1973] 1994: 35). This falls foul of requirement (iv), as the operation of the second hand of a watch does not in any meaningful sense contribute to an activity of the whole watch. Moreover, like Canfield's definition, it falls foul of (iii), as it fails to distinguish between contributions to the whole system that are accidents and those that are really functions.

More generally, Wright notes that earlier accounts have failed to notice requirement number (vi). This, he thinks, explains the difficulty those accounts had in capturing the difference between functions and beneficial side effects. He notes that in everyday life we often use the expressions "What is it for?" and "Why is it there?" to ask the same question. As we saw in the summary of requirement number (vi) above, he thinks we should take this feature of ordinary language seriously, and moreover that the "why" is asking for a *cause*, albeit a cause of a special type. We also, he notes, often use a third expression – "What does it do?" – to ask the same question. If someone asks "What does the heart do?", the chances are that she will not accept the answer "it goes lub-dub".

With this as a clue, Wright gives us his initial formulation. Given all the requirements he wants it to meet, and given the great complexity of some of the earlier attempts, Wright's initial definition is amazingly simple. The function of *X* is *Z* means "*X* is there *because* it does *Z*" (*ibid.*: 39).

A little explanation is in order. In a way, the simplicity of this definition is made possible by a little cheating on Wright's part. All the expressions that he uses – "is there", "does" and "because" – are to be understood in a slightly loose, colloquial sense. One might object that at least Canfield and Beckner were trying to be precise. But Wright's exploiting of the loose colloquial uses of these terms is defensible. By this means he is able to accommodate a number of subtly different types of case, all of which he thinks should be considered cases of function. The expression "*X* does *Z*" is to be here understood as covering even the cases where *Z* never happens. We might say "This button activates the fire extinguisher", even if the fire extinguisher never gets activated. It still seems natural to say that that is what the button does, meaning that *if called on*, or something like that, that is what it would do. We do use the expression "*X* does *Z*" in that sense. Moreover, "*X* does *Z*" does not have to mean that *Z* is an activity of *X*. We say that pumping blood is what the heart does, and it seems right to say that this is an activity of the heart. But we also say that keeping the rain off is what an umbrella does, but it seems wrong to say that this is an activity of the umbrella. Beckner's second stab at the problem went astray here according to Wright, as it did not allow for functions that were not activities. As for "is there":

> "Is there" is straightforward and unproblematic in most contexts, but some illustrations of importantly different ways in which it can be rendered might be helpful. It can mean something like "is where it is", as in, "Keeping food out of the windpipe is the reason the epiglottis is where it is." It can mean: "*C*'s have them," as in "animals have hearts because they pump blood." Or it can mean merely "exists (at all)," as in, "Keeping snow from drifting across roads (and so forth) is why there are snow fences". (*Ibid.*: 39–40)

Although Wright says these renderings are "importantly different", note that the difference is not between functions and something else. All of the examples in this quotation are functions. As for "because", I have already dealt with that in (vi) above.

Let us see how this simple definition is supposed to satisfy the various demands.

(i) The definition only says that the thing is there because it does X; it does not fill that out with, for example, X is there because it was selected in virtue of doing Z, or X is there because someone put it there to do Z. Thus, it fits consciously designed artefacts and evolved traits of organisms equally well.

(ii) In the case of a watch's second hand, it is there because it enables a person to read off seconds more easily, regardless of whether the watch's owner actually uses it during the watch's lifetime. It being easier to read off seconds is a consequence of the second hand's being there, even if the owner does not actually use it. But, because of the gloss that Wright puts on the term "what it does", the second hand does not have to be actually used for that to be its function.

(iii) As for the accidental effect, this is dealt with by the "because". Even if the second hand brushes away bits of dirt, and even if that is extremely useful, that is not the second hand's function unless it was put there in order to do this. As for a biological trait, we know that a trait can have some useful accidental effect (e.g. the bones in a baby's skull that I discussed in Chapter 3). And, presumably, it could eventually come to be selected for because of this effect, and not for the function it originally had. But if this latter happens, only *then* we would say that the trait came to have the new function and not the old one.

(iv) The definition does not say that X has to be necessary for the system of which it is a part to work at all. It applies equally well to functional items of which this is false as to ones of which it is true.

(v) This is slightly more complicated. We have to distinguish between cases where something fails to do what it does because some external circumstance prevents it, and cases where it fails because it is itself defective. The function of chlorophyll is to photosynthesize, but it may fail to do so because the plant is in the dark. This is a different case from: the function of the heart is to pump blood, but it fails to do so because it is defective. The first type of case can be dealt with in the way that Wright dealt with requirement number (ii). Just as the gloss he put on "does" allows us to say that the button activates the fire extinguisher even if it is never used, it also allows us to say that chlorophyll photosynthesizes, even if it is prevented from doing so. Moreover, the definition not only does not say that X actually does Y, but it does not even say that X *is capable of* doing Y. So, paradoxically, we can say that pumping blood is why a person's heart is there, even if it is a defective heart that cannot pump blood. (Further problems are shortly about to arise with this, however.)

(vi) This is captured by the "because".

Unfortunately, things are not quite as simple as all that. Straight away, Wright comes up with a counter-example to his own initial formulation. Oxygen combines readily with haemoglobin, so that is something that oxygen does. Moreover, the fact that oxygen combines readily with haemoglobin explains why it "is there", that is, why it is in our blood. But it seems wrong to say that combining readily with haemoglobin is the *function* of the oxygen in our blood. Presumably, the function of the oxygen in our blood is to take part in chemical reactions in our cells. Wright argues that the difference between combining with haemoglobin and taking part in chemical reactions in our cells is that the latter is a *consequence* of the oxygen being there, whereas the former is not a consequence, even though it is a reason the oxygen is there. Accordingly, he expands his formulation:

> The function of X is Z means:
> (a) X is there because it does Z,
> (b) Z is a consequence (or result) of X's being there.   (*Ibid.*: 42)

For this to work, we need to think of "is a consequence of" in the same way as we are to think of "does". That is, we have to be prepared to say that turning the fire extinguisher on is a consequence of the button's being there, even if the button is never pressed. Even if we accept this, however, this revision creates a problem for the solution to requirement (v). Think again of the faulty heart or the faulty second hand. It would surely be stretching things to say that pumping blood is a consequence of a heart that cannot pump blood being there. As a way of dealing with this, Wright suggests that we think of the word "function" as being in inverted commas when we speak of the function of a faulty item. The function of a faulty heart is not to pump blood. In fact, a faulty heart does not have a function. Rather, it has the "function" of pumping blood, because a normal heart has the function of pumping blood.

7.2 What it has been selected for

As usually happens when philosophers offer definitions, Wright's definition of function has been attacked on the grounds that things that fit the definition are clearly not functions. For example, Christopher Boorse (1976) offers the following scenario. A gas tap is left on in a laboratory. The gas knocks out the only scientist in the laboratory, so he cannot turn it off or open a window, so after a while there is more gas in the laboratory. Knocking out the scientist is why the gas is there – that is, in the air in the laboratory – which seems an acceptable use of "is there" on Wright's construal. Knocking out the scientist is also a *consequence* of the gas being there. So we would have to say, on Wright's definition, that knocking out the scientist is the function of the gas.

Millikan (1984; 1989) proposes getting rid of such problems by making the function of X what X was *selected for*. What she is interested in, she says, is *biological* function. Accordingly, she is not concerned about satisfying Wright's criterion (i), that the concept of function that we apply to parts of living things is the same concept that we apply to man-made artefacts. Her definition of "proper function" is, A has function F as a proper function if *either*:

> (1) A originated as a "reproduction" (to give one example, a copy, or a copy of a copy) of some prior item or items that *due* in part to possession of the properties reproduced, have actually performed F in the past, and A exists because (causally historically because) of this or these performances. [Or] (2) A originated as the product of some prior device that, given its circumstances, had, performance of F as a proper function and that, under those circumstances, normally causes F to be performed by *means* of producing an item like A.   (Millikan 1989: 288)

This formulation does well in terms of the rest of Wright's requirements. It recognizes that a thing's function has a role in explaining why it is there. It allows that something can have a function even if it never performs that function. It does not make usefulness a sufficient condition for something being a function; that is, it distinguishes between genuine functions and mere accidental effects. It does not say that the thing has to be necessary for the system as a whole of which it is a part, and it allows that something may be unable to perform its function. It does all this by making the fact that something has done F the reason for saying its function is F. So effects of biological functions that were not selected for do not count. Moreover, the gas tap example does not count either, as the gas in the room is not there in virtue of being a reproduction of something that had that effect.

Millikan's definition seems to do a good job of accounting for biological function. But we still might be worried about the fact that it no longer works for the functions of man-made artefacts or their parts. Wright wanted a definition of function that would apply equally well to parts of living things and to man-made artefacts. And this does seem to be an attractive feature of his account. Millikan's formulation does not actually mention natural selection, let alone make it a requirement that the entity in question has to be biological. So, in principle, it can apply to man-made artefacts. If we think of mass-produced items (such as cars), or items that human beings have been making for some time (such as ships), we might plausibly say that the reason they have the various functional parts they have is that those parts have worked and hence have been reproduced. On this construal of things, the functional parts of such items can be said to fit Millikan's definition. The steering wheel of your car really did originate as a copy of some prior item (the steering wheels of earlier cars), and has its function in virtue of those predecessors performing that function. Millikan is perfectly happy to say that any man-made items that fit her definition count as having functions. But what about a man-made item that is unique, or the first of its kind? The Wright brothers' first aeroplane clearly had parts that were functional, but they cannot be considered reproductions of parts of anything else that performed the same function. Even if there were earlier prototypes, they did not work. We would probably say that it was something to do with the *intentions* of the Wright brothers that gave those parts the functions they had. But Millikan's account does not allow intention, even intention intelligently carried out and successfully achieved, to count as a reason for calling something a function.

Her justification for taking this stance is that it is *biological* function that she is concerned with. This means, incidentally, that even if the gas tap example had not been dealt with, it would not matter. So whether or not it turns out that some non-biological items wrongly fit or fail to fit the definition, is of no consequence. More generally, she is interested in producing a definition of function that reflects scientific practice, not one that conforms with whatever our everyday intuitions or ordinary language tells us. This tough-minded approach allows her to brush aside another class of counter-examples, those that are based on science-fiction type counter-examples. For example, suppose an atom-by-atom replica of a lion appeared suddenly by a chance configuration of atoms. This would not have the right kind of history for any of its parts to fit Millikan's definition, but we would still want to say that its "heart", "lungs" and so on have functions. For that matter, they would not fit Wright's definition either. However, Millikan argues that there is no need for any definition of function to be required to fit such cases, as this is not a situation that would ever arise in the real world. She would say: none of the instant lion's parts have any functions, but so what? Science is none the worse for working with a definition that conflicts with what ordinary language would say in cases that never arise in the real world. So for Millikan the only requirements that we should require a definition of "function" to satisfy are that it apply to things that could actually exist in the real world, and that it conforms to how scientists actually use the term in the contexts in which they actually use it.

The thought seems plausible if we think of cases like this: suppose scientists have a definition of "liquid" according to which glass is a liquid. (In actual fact, glass is not clearly either a solid or a liquid according to physicists.) A philosopher might then object: in ordinary language we do not call glass a liquid, so the definition must be wrong. But a scientist could respond that we need a definition of "liquid" for scientific purposes, and the definition that makes glass into a liquid is good enough for that. Moreover, if we wait around for a definition that fits our ordinary-language usage perfectly, we may be waiting a long time. Philosophers are very clever at thinking up counter-examples: examples of things that fit a proffered definition of an X but are clearly not X's according to ordinary language. To get around this, Millikan proposes that we say, in effect, to hell with ordinary language. For scientific purposes, we do not need to deal with instantaneous lions, so whether they have functional parts or not according to this definition is irrelevant. The selected-effect definition of function, then, is perfectly adequate for scientific purposes, and to demand more of it is unreasonable:

> Now I firmly believe that "conceptual analysis", taken as a search for necessary and sufficient conditions for the application of terms, or as a search for criteria for application by reference to which a term has the *meaning* it has, is a confused program, a philosophical chimera, a squaring of the circle, the misconceived child of a mistaken view of the nature of language and thought. (Millikan 1989: 290)

In my opinion, however, this is all too quick. Millikan may well be right that ordinary language should not be treated as the arbiter of whether a definition that is to be used in science is good enough for science. And she may be right that a definition that deals with all the counter-examples that philosophers can dream up will probably be a long time coming. But scientific purposes are not the only purposes we have in trying to define something. Moreover, saying that man-made artefacts are for the jobs they are designed to do hardly seems like a crazy philosopher's counter-example. It is not clear that, in deciding if a definition is adequate, we are faced with the choice of *either* deciding on the basis of whether it fits what ordinary language would say about the counter-examples, or deciding on the basis of whether it is good enough for science. Millikan appears to think that if we are not willing to accept a definition that is "good enough for science" then we must be unreconstructed ordinary-language philosophers. But that presupposes that these are the only two possible grounds we might have for accepting or rejecting a definition.

Imagine that someone came up with a definition of "alive", and it turned out that light bulbs fit the definition. A philosopher produces the counter-example of light bulbs to show that there is something wrong with the definition. But in doing so, that philosopher is not necessarily arguing that there is something wrong with the definition *because it conflicts with ordinary language*. It is possible that the philosopher is arguing that there is something wrong with the definition *because light bulbs are not, in fact, alive.*

This might seem a silly example, indeed, another case of the philosopher's disease of thinking up highly unlikely scenarios to create problems that never arise in practice. Is it remotely plausible, it might be asked, that a definition of life might be produced that would fit light bulbs? Is this not just as unlikely as instantaneously appearing lions? Moreover, we know that science has sometimes given us good reason to reclassify things. Who would now think it reasonable to doubt the classification of earth as a planet, for example? Yet that classification came about as a result of a revolution in science. So, this argument might go, if the purposes of science require us to move things into, or out of, categories contrary to everyday usage, so much the worse for everyday usage. It would have been no good for a seventeenth-century astronomer to argue that we cannot reclassify earth as a planet *because it is not a planet*. We would need, at least, to show that there was some *reason* for saying it is not a planet, other than ordinary usage, and that that reason is sufficiently important to outweigh the scientists' reasons for calling it a planet. By analogy, we would need some reason for accepting that some things that fulfil the definition of function are not actually functions, or vice versa, other than in ordinary usage. That reason would have to be sufficiently important to outweigh the consideration of scientific practice. Can we do this?

I think we can. The notion of function, even within science, is not confined to biological function. *Engineering* may be considered a part of science, but engineers will talk of the parts of things having functions. Moreover, the way engineers talk about function in machines is clearly the same way that biologists talk about function in body parts. The outcome of a part being there, is in both cases to be accounted for by the same physical laws. Biologists may even borrow information from engineering to understand how an organ works. The same principles of aerodynamics apply to a bird flying as to an aeroplane flying, even though they fly using different mechanisms. And an engineer may study a previously unseen machine to discover its function – for example, in industrial espionage – in just the same way that an anatomist may study a previously unseen organ to discover its function. Dennett (1995) calls both processes "reverse engineering". The machine–organism analogy, and specifically its "what-is-for-what" aspect has a role even in science. It may not be a matter for concern that the parts of instantaneous lions turn out not to have functions on this definition. But it is a matter for concern that the parts of the Wright brothers' first aeroplane turn out not to have functions either.

In any event, the appeal to scientific practice as the arbiter has led some philosophers to challenge accounts of function that focus on the purposeful or "for something" aspect.

7.3 What it does

A radically different account of function comes from Cummins ([1975] 1994). Like Millikan, Cummins is concerned to come up with an account of function that harmonizes with the term's usage in scientific practice. Unlike Millikan, he does not make a thing's having been selected to do something a criterion for that something being its function. His grounds for doing this are that it is, in fact, often extremely difficult to reconstruct the evolutionary history of a trait. In many cases we do not know anything about its evolutionary history, yet we can still say that it has a function, and what that function is. This might sound like arguing that people knew what gold was before they knew that its atomic number is 79, so therefore having the atomic number 79 cannot be the definition of gold. Millikan explicitly rejects such arguments, arguing that the definition of gold is: the element with atomic number 79. The development of science, she points out, has sometimes led us to take apart what we thought were natural kinds because by scientific criteria they did not constitute properly defined classes. What was once called consumption turned out to be a number of different diseases. The fact that people identified it as consumption before they knew anything about the causes of the different diseases does not show that the concept of consumption needs to be defined without reference to those causes. It shows that the people who thought consumption was one disease were wrong. Similarly, Millikan argues, the fact that we can identify functions without knowing the evolutionary history does not show that the concept of function can be defined without reference to evolutionary history. It shows that someone who assigns a function where there is not the right kind of evolutionary history is wrong.

However, Cummins is not slipping back into what Millikan sees as the vice of ordinary-language philosophy, or conceptual analysis. His point is that *scientists* can identify functions without knowing the evolutionary history. It is a large part of the business of anatomy to identify the functions of various parts of the body. The function of valves in veins is to close up if blood temporarily swashes the wrong way, and this was discovered by William Harvey long before the theory of evolution came along. Hence, Cummins's account is not an appeal to the fact that some things just intuitively seem to be correctly called functions and others not. Like Millikan's, it is an appeal to science.

Nor is his account friendly to Wright's. Wright makes the function of something the reason it is there. In the case of biological traits, divine explanations aside, a thing's function is the reason it is there if and only if it was selected for its ability to perform that function. So, we could only say that doing X is the reason Y is there if we knew the relevant evolutionary history. Once again, Cummins would say, Wright's formulation fails to take into account the fact that scientists seem to manage to identify what something's function is without knowing its history.

Cummins eschews any attempt to capture the notion of function as "what it is for". Instead, he proposes to explicate it in terms of *what it does*. He notes that functional explanation is a distinctive type of explanation, one that pervades biology through and through. In particular, it is different from explanations in physics, which are in terms of laws. Moreover, not everything is subject to functional explanation. Orbiting the earth is what the moon does, but we do not say that that is its function.

To explain his account, Cummins appeals to the notion of a system of parts, which collectively produce some effect. The standard biological examples will do. The eye has parts that collectively have the effect of transmitting information of a particular kind to the brain. The heart has parts that collectively have the effect of pumping blood around the body. Now think of how we might explain how the eye produces the effect it does. Our explanation might say things such as: the lens focuses light; the iris expands and contracts in response to light levels; and so on. Note that this explanation makes no reference to physical laws: it only refers to effects that those parts of the eye produce. Nonetheless, they are not explanatorily empty; they do genuinely explain how the eye works, in as much as, when we have said what all the different parts do, in the sense of what effects they produce, we have explained how the eye does what it does. Note also that such a functional explanation can be taken further. Say we want to explain how the iris does what it does. We could give an explanation in terms of *its* parts and what *they* do. In other words, we *functionally analyse* it. And so on until *eventually* we get to parts where we can explain what they do straight-forwardly in terms of physical laws.

If and when we eventually get to that stage, we are no longer doing functional analysis, but physical explanation. Distinctively functional explanation picks out what a thing does, and how that in turn contributes to a system of which it is a part. As with the other accounts, we have to take into account that a thing might never actually perform its function. Accordingly, Cummins stresses that his account is a *dispositional* account: it is sufficient if the thing is disposed to do the thing that counts as its function, just as a glass is disposed to break if it is hit with a hammer. Likewise, it is what the system as a whole is disposed to do that is explained by what the parts are disposed to do. Here is his formulation:

> x functions as a φ in s (or: the function of x in s is to φ) relative to an analytical account A of s's capacity to ψ just in case x is capable of φ-ing in s and A appropriately and adequately accounts for s's capacity to ψ by, in part, appealing to the capacity of x to φ in s.
>
> (Cummins [1975] 1994: 64)

The above example should help in understanding this. An "analytical account of a system's capacity" is an account where what a system as a whole does is analysed into what its parts do, as in my (very partial) analysis of the eye. "Appropriately and adequately accounts for" means fully accounts for, as a complete version of my analysis of the eye would presumably do.

There are a few other nuances to Cummins's account, but I shall mention just one, very important, one. Astute readers may have noticed that, for all I have said so far, the various parts of a rock rolling down a hill would count as having functions. We could analyse what the rock does into what its parts do. The left half is rolling down the hill, the right half is rolling down the hill, and when you put the two together you get the whole rock rolling down the hill. One might try to counter this counter-example by saying: but that is not an *explanation*. This would be correct, but we need to say something about why it is not. With objections like this in mind, Cummins adds the proviso that an account only constitutes an *analytical* account if what the parts do is *different in kind* from what the whole thing does. Thus the iris counts as having a function as part of the eye, but the parts of a rock do not count as having functions as parts of the rock.

On this account, a thing's function is indeed something it does, just as on Wright's account or on option (1) of Millikan's. Moreover, it is not just any old thing it does, but an effect it produces, again just as on Wright's account, and Millikan's option (1). Moreover, Cummins's account fits many of the same cases where we would say "function of X is Y", as these rival accounts do. The iris is for reducing or increasing the amount of light getting into the retina, just as the other two accounts would say.

Cummins's account, unlike Millikan's, captures one of the aspects of function that make Wright's account appealing. The word "function" means the same thing whether we are talking about a man-made artefact or a biological trait. Maybe, however, it does not capture this aspect as well as Wright's account. It is not clear how it would apply to very simple man-made artefacts. The function of an umbrella is to keep off rain and we can say that keeping off rain is a dispositional property. But there does not seem to be a bigger system, of which the umbrella is a part, and the operation of which is explained by functional analysis with the umbrella as one of the components.

However, there is a more serious problem than that. Just as I argued that Millikan's account failed because it lost sight of the concept of function as being common to biological traits and artefacts, Cummins's account fails because it loses sight of the concept or function both as involving purposefulness and as explaining why a thing is there.

Let us return to Cummins's reason for rejecting accounts of function that attempt to capture the "being for something" aspect, which include both Wright's and Millikan's accounts. Cummins's argument is that we often do not know the evolutionary history of something, yet we still know its function. But Millikan's claim is not that we have to *know* the evolutionary history of something for something to have a function, but just that something has to *have* the right kind of evolutionary history in order to have a function. Gold would still be whatever has the atomic number 79, even if we had extreme difficulty in identifying things as having the atomic number 79. But, Cummins might reply, we want an account of function that harmonizes with scientific practice. We want an account of function that allows us to say that scientists actually successfully identify functions, because they do. This reply would not work, however. Millikan can reply that we do not need to know the evolutionary history of something in order to know, at least with a pretty high degree of certainty, what it was selected for. Think of an intricate mechanism, as so many biological mechanisms are. The eye, the heart and the leaves of plants are all composed of a complex, highly organized system of interacting parts that, working together, produce effects that are highly useful, even crucial, for the organisms of which they are a part. The only remotely plausible explanation for this state of affairs is that these parts were selected for those functions, maybe not throughout their entire evolutionary history, but through a long enough stretch of it to have made them well designed for what they do so well now. The lung may be a modified swim bladder, but it is clearly very well designed as a lung, so it is reasonable to conclude that it has been selected for, and is still there because of, its ability to function as a lung. So, *contra* Cummins, we can know that it was selected for this, without knowing its evolutionary history.

There are, of course, times where we wrongly think that something was selected for something. But in such cases, we are also wrong about that thing being its function. There are also times where something is a borderline case, perhaps because it is not a very complex design, or because it has only recently been co-opted for its current purpose. But that is also a borderline case of something having something as a function. These problem cases leave the core cases – eyes, hearts and many more – untouched.

7.4 Conclusion

A puzzling feature of Cummins's account is that he sees it as a *rival* to those that seek to capture the "what it is for" aspect of function. It is plausible to think of Wright's and Millikan's accounts as rivals to each other, as they both seek to capture the "what it is for" aspect, but disagree as to how it should be captured. But Cummins's account sets out to capture something different. Moreover, Cummins's account captures a genuinely important concept in biological explanation. And, yes, that concept does go under the name "function". A tempting conclusion would be to say that Cummins is just talking about something different from the others, even if it happens to have the same name. Certainly, I would not want to be without the concept that he is talking about, or his useful explication of it.

But that is not the end of the story. Finally, I want to say that the word "function" has a life outside biology: indeed, outside science in general. We ask "What is the function of the corner flags in soccer?", meaning: why are they there? It would strain either Cummins's or Millikan's account of function to say that their function is to make it easier for the referee to make decisions: Cummins's because that is a purpose and Millikan's because the first ever corner flags presumably had that function. Wright's account just seems more *natural* than Millikan's or Cummins's accounts. In case that does not cut any ice, as it presumably would not with Millikan, I add that Wright's account does not do noticeably *worse* than Millikan's as an account of precisely that concept of function that scientists use, despite the fact that that does not seem to have been Wright's aim, and despite the fact that his account is considerably simpler (itself not an unworthy consideration). We can then think of the further question "How does it do as an account of our more everyday use of the word?" as the tie-breaker. In this case, Wright's account wins. Speaking personally, if I wanted to convey to my ten-year-old niece that the function of the inner-ear canals is to help us balance, I would not say that that was their selected effect, still less give a Cummins-style analysis. I would use Wright's original simple formulation: they are there because they help us balance.

8. Biological categories

8.1 Introduction: natural kinds in general

One of the salient facts about biology that is explained by evolution is the existence of a branching tree of life: a set of categories in which every organism finds its proper place. These placings are not merely matters of human choice or convenience, but are dictated by facts about the organisms themselves. For example, it would be *wrong* to classify a whale as a fish. This suggests that the biological categories species, genus and so on are prime examples of what are known as *natural kinds*. Put briefly, natural kinds are categories that are actually there *in nature*, as opposed to being impositions on nature for our own convenience. It is often said that natural kinds are the categories that are of interest to science. For example, Kripke (1980) holds that it is the business of science to discover natural kinds.

A prime example of a natural kind is a chemical element. All samples of, for example, gold have the same atomic number – 79 – and it is a central pillar of the science of chemistry that elements can be assigned to their proper place in the periodic table of the elements. An alloy is a mixture of two or more metals, but that does not mean that the boundary between one metal and another is vague or fuzzy. A zinc–copper alloy, for example, is not made of atoms that are partly zinc and partly copper; it is made of atoms that are zinc and atoms that are copper.

Moreover, we can group elements together based on features that are of scientific interest. For example, the alkali metals – lithium, sodium and so on – are grouped together because their atoms have one electron in the outer shell. This common underlying feature can explain observable properties that the alkali metals have in common: the ways in which they can combine with other elements. "Alkali metal" can thus be considered a natural kind. On the other hand, a category such as "used in making coins" picks out a certain group of metals – gold, silver and so on – but it is clearly not a natural kind. The fact of being used in making coins is an *accidental* feature of gold. Gold would still be gold even if it was not used in making coins. This fact about gold is of no scientific interest. Properties that something must have if it is an example of a certain natural kind are *essential* properties of that kind. It is, in this sense, an essential property of some substances that they dissolve in water. But this is a feature of the substance that is explained by something more fundamental: its atomic or molecular structure. Thus, while a property such as "dissolves in water" is essential, and is of scientific interest, it is of less fundamental scientific interest than the atomic or molecular structure. We could consider the category "substances that dissolve in water" to be a natural kind, but its status as such would be of subordinate importance to the categories "gold", "alkali metal" and so on.

8.2 Taxonomy

Taxonomy is derived from the Greek *taxis*, which can be translated as "order" or "arrangement". The Greek word can refer, for instance, to the battle array of an army, which gives us a clue to what is striking about the order that we call "taxonomy" in biology. An army is divided into military units: divisions, platoons, companies and so on. We could, if we chose, classify the soldiers in other ways: into those over and those under six feet tall; into those with blue eyes and those with brown eyes; and so on. However, for most military purposes – for determining who is answerable to whom, for example – the division into military units is the one that matters. Similarly, when it comes to the biological world, certain classifications are important and others are unimportant. We might be tempted to classify whales along with fish, or bats along with birds, on the grounds of the means of transportation they use, or to classify whales along with dinosaurs on grounds of size. But biology tells us, loud and clear, that we would be wrong.

We have become used to thinking of species as descended from earlier species, so we now accept that there is a good reason for insisting that some biological classifications are genuine and others are spurious. In *Moby-Dick* Herman Melville takes issue – with what level of seriousness it is difficult to tell – with Linnaeus's separation of whales from fish:

> In his *System of Nature*, AD 1776, Linnaeus declares, "I hereby separate the whales from the fish." But of my own knowledge, I know that down to the year 1850, sharks and shad, alewives and herring, against Linnaeus's express edict, were still found dividing the possession of the same seas with the Leviathan.   (Melville [1851] 1972: ch. xxxii, 230)

Melville goes on to weigh Linnaeus's criteria for classification against those of his whalemen acquaintances, who call whales fish, and professes to find the latter more persuasive. There is a serious question behind all this: *before* we accepted the descent of species from other ones, what were the criteria for accepting one classification rather than another as *the* genuine one?

It is striking that, even long before Darwin, the impulse to determine the genuine scheme of biological classification was strong. It begins in Aristotle's biological writings, and can be traced through to the aforementioned Linnaeus and beyond. Not being a historian of science, I am unable to give a comprehensive answer to why this was. Doubtless it was partly for philosophical reasons, but it may also have a basis in our pre-philosophical encounters with the phenomena themselves. Linnaeus's grounds for classifying whales as mammals are "their warm bilocular heart, their lungs, their movable eyelids, their hollow ears, penem intrantem feminam mammis lactantem" (*ibid.*). It is not just *one* criterion that justifies the classification, but a cluster of them. What we find again and again in nature are groups of traits that tend to be found *all together* in groups of organisms. Birds tend to have feathers *and* beaks, for example. The words "tend to" are an important caveat here. We might still classify a creature in group G despite its not having all the traits possessed by most of the creatures in group G. (This element of vagueness in biology raises important questions, which I shall address in Chapter 10.) The division of organisms into groups is not a division by means of walls that are at all points unbreachable. Rather it is a clustering together of collections of traits, as an area of country might be more densely populated in some parts than others. However, this clustering together is very strongly marked, as if there were some areas with the population density of Hong Kong and others with that of the Sahara Desert.

In the system of classification that may be said to have reached maturity with Linnaeus, biological units are, like the division of an army into military units, *nested*. That is, bigger units are divided into smaller units: kingdom → phylum → class → order → family → genus → species. There are no overlaps between different groups. A class cannot belong to more than one phylum, for example. As with teleology, so also with taxonomy; the more biology progressed, the more sheer complexity was uncovered. The Linnaean system employs seven levels of classification. Moreover, the system once again exhibits order *over and above* the order uncovered by the physical sciences. The latter do contain their own grounds for classifying things: H_2O, CO_2, $NaCl$, acids and alkalis, inert gases and electricity conductors are all classifications given by the physical sciences. But the physical sciences by themselves give us no reason to expect there to be classifications of the kind we find in biology at all. Why, for all physics tells us, should creatures with feathers also have beaks? Moreover, the physical sciences give us no grounds for taking some biological classifications to be genuine and others to be spurious. Classifying whales along with fish because their preferred mode of transport is swimming would, as far as the physical sciences are concerned, be perfectly fine. If we insist on holding on to our conviction that some biological classifications are genuine and some are spurious, the physical sciences do not appear to be able to explain why the genuine classifications exist.

Finally, the taxonomy uncovered by scientists over the centuries does not seem to be an artefact of their own making. To say this is to disagree with Locke in *An Essay Concerning Human Understanding*, where he writes: "Since the Composition of those complex *Ideas* are, in several men, very different: and therefore, that these Boundaries of *Species*, are as Men, not as Nature makes them, if at least there are in nature any prefixed Bounds" ([1690] 1979: bk III, ch. VI, § 30). Despite the use of the word "species", Locke was not referring specifically to *biological* kinds, but they are to be taken as included. More recently, some philosophers are fond of saying that we – or our language, which is sometimes made to sound as if it has an inexorable control over us – can "carve up" the world in various different ways. For example, we could have a word, "teeble", for green tables, and another word, "towble", for brown tables, and might think that that classification was more significant than a classification of things into tables and non-tables.[1] For that matter, we could have a word that referred to the bottom half of a cup and the part of the table defined by the six-inch radius around the cup, and think of the entities so referred to as perfectly respectable denizens of the world, while not thinking that either cups or tables are.

Whatever the merits of this view in general, if we applied it to biology then we would have to accept that calling a whale a fish was just as good as calling a whale a mammal. And this, in turn, is gainsaid by the conviction of many people, from Aristotle to Linnaeus and beyond, that there is a *genuine* way in which the biological world is carved up. Seeking out biological classifications may be just a habit, but it is a deeply ingrained habit. Very young children seem to find it perfectly natural to distinguish human beings from dogs, and dogs from cats, and taxonomists have been plying their trade for millennia. Moreover, if the claim applied to biological classification is – as it appears it would be – that classifying whales along with fish is *just as good* as classifying them along with mammals, then it fails to take into the account the strongly marked clustering together of traits that we actually find. Locke's argument that "those complex *Ideas* are, in several men, very different" seems to be based on viewing things in a very all-or-nothing way: the word "dog" might conjure up a slightly different cluster of ideas to me than it does to you, but there will be significant overlap between our respective clusters. In the overwhelming majority of cases two people can agree on whether something is a dog or not. So we have good reason to believe that the taxonomy of living things is a genuine feature of the world. Moreover, like teleology, it exhibits an order over and above that uncovered by the physical sciences.

8.3 What are the natural kinds of biology?

Linnaeus's classifications match up, in many instances, with the taxonomic classifications used today. We still use the classifications "bird", "mammal" and "fish", and, despite Melville, whales are still classed as mammals. The Linnaean system did reasonably well when it comes to those parts of the tree of life that are close to us; the old subdivisions of mammals stand up well, and those of the chordates only a little less well. Among invertebrates – that is, the whole animal kingdom apart from vertebrates, which do not even make up a whole phylum – biologists in the eighteenth and nineteenth centuries made far too few subdivisions. Groupings that we would now recognize as of at least the same rank as Chordata, were considered to be of lower rank. That is, what we now consider whole phyla were regarded as lesser groupings, such as orders. Moreover, at one time there were just two recognized biological kingdoms: plants and animals. But fungi are now considered a kingdom in their own right, and our knowledge of single-celled organisms has expanded enormously, giving rise to a whole host of newly named taxonomic classifications at the kingdom level.

How, then, are we to decide on the correct way to classify organisms? There are two different questions here:

(i) *How are we to decide which creatures should be grouped together?* Do A and B belong to a group that excludes C, or do A and C belong together in a group that excludes B, and so on? For example, why should we put whales and human beings in a group (mammals) that excludes salmon? When we discover a new type of creature, how do we know what group to put it into? Should the "cat-fox" – the creature that was recently photographed for the first time in Borneo – be classified along with cats or with foxes (or, as is more likely, with neither)?

(ii) *How are we to decide how to rank different groupings?* Supposing we agree that creatures A, B, C, … form a natural group of some kind. How are we to decide if that group should be considered a phylum, a class, an order, or what have you? For example, why are fungi now called a kingdom? How are we to decide if two different groups are of equal ranking? Why are Chordata and Arthropoda (the group that includes insects, spiders, and crustaceans) both considered phyla? Why are insects not a phylum, given that, as we are always being told, there are so many of them?

8.3.1 Phenetics versus cladistics

One way we might try to settle these questions is by simply sticking to the old Linnaean system of classification. Probably no one would suggest this today. For one thing, it would give us no guidance on how to classify those huge areas of the living world that were unknown in Linnaeus's time. Moreover, in practice, creatures have been repositioned on the tree of life when there seemed to be good reason to do so. A milder suggestion would be to at least give the traditional subdivisions some weight, that is, *only* revise them when we have a good reason to do so. But this immediately suggests the question: what, exactly, is a good reason?

Two basic approaches to the first of our two questions have been suggested. The first, called phenetic taxonomy, is based on how similar organisms are to each other. The second, called cladistic taxonomy, is based on what creatures are descended from what.

The phenetic approach has been defended by Sneath and Sokal (1973). One thing it has going for it is that it appears to classify biological kinds in the same way that we classify chemical elements, those exemplary natural kinds. The division of chemical substances into natural kinds is based on features that the substances *currently* possess. An atom of gold currently has the atomic number 79. Were that to change, it would no longer be gold. Before the theory of evolution came along, biological classifications were similarly based on current features. Linnaeus classified whales as mammals because they possess mammary glands, warm bilocular heart and so on. In making this decision, he probably thought of these as essential features of mammals, and having legs or living on land as non-essential. What happens, however, if two biologists disagree on what the essential features of some biological kind are? Suppose, for example, that one of them decided that being fish-shaped made something a fish? We might end up in a situation where the classification of whales was ambiguous, or a matter of the individual biologist's preference. But then the intuition that there is a *right* way to classify organisms would be frustrated. Moreover, biology, not being able to discover genuine natural kinds, would, at least on Kripke's view of things, fail to be a science.

We could, if we liked, divide the world of living things up in many different ways, based on features that organisms genuinely possess. For example, we might decide that *vertebrates that swim* or *warm-blooded animals that fly* were biological kinds. The first category would include fish, whales and others; the second would include birds and bats. There would be some scientific interest in these categorizations. Being a vertebrate that swims would undoubtedly carry with it some distinctive adaptive challenges, ones that are not faced by vertebrates that do not swim or by invertebrates that do. They would not be *irrelevant* to science in the way that "being used in making coins" is. But they would be of less fundamental scientific interest than the classification that puts whales with mammals and not with fish. Perhaps they would be analogous to the classification of chemical substances "dissolves in water".

However, it would be highly misleading to consider phenetic classification to be arbitrary. Copi's suggestion is that we use the classifications "teeble" and "towble" just as easily as we now use the classification "table". So it is tempting to think that we could similarly "easily" divide living things up into categories that are completely at odds with the ways that biologists actually do divide them up. We could, if we wanted, categorize creatures into "more than twenty feet long" and "less than twenty feet long". The former category would include whales, many trees, but no human beings, dogs or amoebae. Such a category would not be *completely* vacuous in terms of implications – for example, the ability to pass through certain openings – but, actually, it would not include newborn whales or young trees. So it would require us *either* to have some creatures change category during their lifetime, or to modify the criterion to something like "twenty feet long in adult life". But if we did the latter we would still have the problem that there might be individual whales that do not grow to the normal length of their kind, for whatever reason. To get around this, we could simply accept that some individual whales belong to one grade and some to another, but this would cut right across what are, both intuitively and for scientific reasons, very plausible ways to divide up the living world. (As also, of course, would the "solution" that involved individual creatures changing category during their lifetime.)

Of course biological classification is nothing like this, and never was. There is a very good reason why phenetic classification is not nearly as arbitrary as my extrapolation of the Copi argument suggests. The division of tables into "teebles" and "towbles" is based on *one* characteristic only: their colour. Thus the category "towble" would include big brown tables and small brown tables, brown tables made of wood and ones made of plastic, tables that are painted brown and ones that are brown all the way through. By contrast, imagine that brown tables were found to have many other things in common with each other. Imagine that an overwhelming majority of them were roughly the same size, that an overwhelming majority of them were made of wood, and so on. This would mean that, once we know something was a towble, we would be able to reliably (although not infallibly) infer many other things about it: that it is likely to be made of wood and so on. Imagine further that there were some *green* tables that had most of the other properties that brown tables typically possess: that they are the same size, that they are made of wood and so on. It might turn out that some green tables had much more in common with most brown tables than with most green tables. So it might make sense to classify them as towbles rather than teebles despite their colour. Moreover, it would be much more useful to do so, as most of the things that were true of most towbles would also be true of these particular green ones. Thus, most inferences we made about them on the assumption that they were teebles would be false, whereas most of the inferences that we made based on the assumption that they were towbles would be true. Note the occurrences of "most" and "nearly the same size" and so on in all this. There need not be any property that all and only towbles have in common, not even the property of being brown tables. Rather, there could be a large *cluster* of properties that any one towble had most of. This is, in fact, the situation that we find with groups of living things. Phenetic classification can be based on large clusters of properties. Even if Linnaeus thought that some properties of mammals were essential, it is possible to do phenetic classification without this assumption. In practice, it is based on as many properties as we can practically measure (see Quicke 1993). This is why Linnaean classifications were not arbitrary. Pre-Darwinian biologists were able to point out large clusters of similar properties in groups of organisms, and base their classifications on them.

But recall why I said that the classification "dissolves in water" is less fundamental than the classification by atomic number. It is that the property picked out by atomic number *explains* why something dissolves in water, *and not the other way round.* And this explanation is linked to the fundamental theory that unifies the facts of chemistry: atomic theory. But what is the fundamental theory that explains the features of living things, and unifies the facts of biology? The answer, I hope, is clear by now: the theory of evolution.

Thus, the classifications that are of fundamental interest to biology are ones based on evolution. More precisely, it is the evolutionary history of a creature that provides a unified explanation of its properties. This includes both properties that are adaptations, and ones that are spandrels. So we should classify together organisms that have a common evolutionary history.

The most popular view among biologists, originating in the work of Willi Hennig (1966), is that the biological classifications that are of fundamental scientific importance are based on *common ancestry*. A group of organisms that consists of all and only the descendants of some common ancestor is called a *monophyletic* group, or a *clade*. A classification based on some other feature is called a *grade*. Biological taxonomy is far more concerned with putting creatures in the right clades than it is with putting them in the right grades. Whales belong with mammals and not with fish because there is a common ancestor that all and only all the other mammals, plus whales, have. To find a common ancestor with whales and fish, you would have to go a lot further back in evolutionary history. And that common ancestor would also be an ancestor of all mammals, not to mention of amphibians, birds and reptiles. For that matter, the reason whales themselves constitute a natural kind, is that they share a common ancestor with each other that they do not share with anything else.

We could classify creatures into those that can fly and those that cannot. Knowing this about a creature would be of some scientific interest, allowing us to reasonably infer things about it, for example, that it has some means of navigating in three dimensions and that it has a body structure that minimizes weight. But assigning a creature to its proper clade usually allows us to infer much more. Knowing that a whale is a mammal allows us to infer that it *probably* suckles its young, that it probably has a vertebral column and a dorsal–ventral main nerve, that it is probably bilaterally symmetrical, that it probably reproduces sexually, that its cells probably contain mitochondria but not chloroplasts, and so on. Most of the items on this list are ones that mammals have in common with other creatures. But in each case those creatures form a clade that includes mammals. The dorsal nerve is a trait that belongs to chordates, which form a phylum. Mitochondria are characteristic of eukaryotes, a "super-kingdom" that includes all animals, plants and fungi. Note well the repeated appearance of "probably" in the list. The inferences are not absolutely secure. Not all chordates are bilaterally symmetrical. Fiddler crabs and some flatfish are obviously not, and even we are not entirely. Our livers and hearts are asymmetrical, for example. However, such clade-based inferences have a high degree of reliability. *Most* chordates are *more-or-less* symmetrical. And this reliability gets higher the deeper into a creature's structure one goes. In fact, there are some creatures descended from eukaryotes that lack mitochondria, such as the single-celled intestinal parasite *Giardia intestinalis*. However, this is extremely unusual, and even *Giardia* has some remnants of mitochondria.

Cladistic classifications focus our attention not just on traits that organisms have in common, but on the *common reason* why they have those traits: the reason of common descent. One might want to grade-classify creatures for a trait they possess in common, on the assumption that they possess it for a common adaptive reason. Fish and whales have similar body shapes for the same reason. But, even without going to the extremes of Gould and Lewontin, it is clear that there is not always just *one* adaptive reason why a creature has a trait. Traits often have dual functions. However, a creature's ancestry is what it is. (We might be tempted to say that no creature has a dual ancestry, but, as we shall see in Chapter 9, that is not true.) In every case a creature has a unique place on the tree of life, a fact that confers on cladistic categories a robustness and scientific interest that grade categories do not have.

But that means that what category a creature belongs to is based not on features it currently possesses, but on its ancestry. Of course, it is largely by means of features that it currently possesses (its anatomy and its DNA) that we decide what its ancestry is, and hence how to classify it. But that does not mean that it is *in virtue of* current features that it belongs in one category or another. Current features are *evidence* of ancestry, but it is in virtue of the ancestry itself that a creature is classified. Clearly, even if a creature was to lose one of its current features it would still have the same ancestry, and so still belong to the same clade. And this is true of *any* feature it currently possesses. I already mentioned the asymmetrical chordates. Even more surprising is the case of rotifers. They are indisputably animals, and hence belong to a clade that is almost entirely sexually reproducing. But they themselves reproduce asexually. There is no move to take them out of the animal kingdom because of this. Even if some creature descended from eukaryotes completely lost its mitochondria, it would still belong with the eukaryotes. Hence, no feature a creature currently possesses is essential to the kind it belongs to. This is a very different situation from that with chemical kinds such as gold.

However, not all the problems that arise in connection with the phenetic approach to taxonomy are eliminated by the cladistic approach, as we shall see.

8.3.2 Biological kinds are vague

One might be tempted to think that the reason grades are not the categories of fundamental scientific interest is that their definitions are rather vague. Do we count as flyers those creatures that glide, such as "flying" fish or "flying" squirrels, or creatures that use their wings to assist long jumps, such as chickens? Does penguins' use of their wings under water count as flying? While there are creatures that *clearly* fly and other that *clearly* do not, there is a large grey area in between. We could, if we wished, sharpen up our criteria for inclusion in a grade. Perhaps we could define "flyer" as something that could, without wind assistance, fly for fifty metres or more (thus excluding chickens) through the air (thus excluding penguins). But such a classification would be drawing a line through what is in fact a continuum. It would be of as much scientific interest as drawing a line between solid objects that are more than five inches long and ones that are not. It would indeed be a genuine distinction, and would probably allow some genuine inferences to be made, such as whether or not the object can be passed through holes of various sizes, but it would not be of any more significance than a division between solid objects that are four inches long and those that are not, or between those that are four and a half inches long and those that are not, and so on. This is because it is a continuum, unlike the division between gold and another element.

Unfortunately, cladistic categorizations are not free of vagueness. The vagueness is of a different kind from that found in grade classifications, however. When it comes to a group of creatures at a single time, either they are a clade or they are not. Fish and whales are not a clade, because there is no creature that is the common ancestor of only fish and whales. Our nearest relations are chimpanzees and bonobos, but there is no currently existing creature that is ambiguous between being a human and a chimpanzee or a bonobo. But if we trace the ancestry of any two living creatures, at some point we will come to a common ancestor. At some point, then, that common ancestor's descendants branched into two different lineages: call them Xs and Ys. The problem is, at what point do we decide to say: these here are Xs, those there are Ys? If we push this thought to its logical conclusion, we get to two individuals, siblings, one who is to become the ancestor of all the Xs, and the other of none of the Xs. We could decide that the immediate offspring of one of those siblings are all Xs and those of the other all Ys. But these first cousins are likely to be *very similar to each other.* They would certainly be classified as the same species. But the *X*s and the *Y*s may be different *phyla*. Classifying two individuals as different phyla just because their respective ancestors later turned out to be different phyla seems a consequence of the definition of clades as all and only the descendants of a common ancestor.

A third possible approach that combines the first two has been suggested. This is known as the *evolutionary* approach.[2] This view was in fact Darwin's own ([1859] 1968: ch. XIII). Because a cladistic approach does not always yield clear answers, it is suggested that we combine cladistic and phenetic criteria; we can do this by treating the fact of common descent as one of the properties in the cluster that we use to decide how to classify something. We might decide to give it especially high weight, but we would not treat it as the sole criterion, or as overriding all others all the time. Thus, two creatures that were destined to become the ancestors of two different phyla could still be classed as belonging to the same species. So the evolutionary approach to classification solves some of the problems of cladism. However, Mark Ridley criticizes both phenetic and evolutionary classification for not being "natural and objective", as he thinks cladism is (Ridley 2003a: ch. 16). The problem is that whereas cladism uses just one criterion, phenetic and evolutionary classification use several. Moreover, in the case of the evolutionary approach those different criteria are *incommensurable*, that is, there seems to be no way of deciding how much weight we should give to the different criteria. Common descent is given some weight, but how exactly are we to decide when it outweighs other criteria and when it does not? What we want is a system of classification based on something in evolutionary history that yields an unambiguous answer to where things should go. The cladistic criterion, when it works, does this: it is absolutely clear that we and chimpanzees are different groups by the cladistic criterion. We may not always *know* whether a group is monophyletic or not, but there is a fact of the matter about whether it is. In that sense, the cladistic approach is objective. So perhaps we should stick to the cladistic approach when it works and only appeal to other criteria when the cladistic approach breaks down. If we chose to be more puritanical than this, we could decide to only ever use the cladistic approach. This would mean that we would sometimes have to accept indeterminate answers. But that seems to be the case no matter what approach we choose.

8.3.3 The ranking question

The reason we do not find ourselves troubled by whether a creature is in phylum *X* or phylum *Y* at the present time is that evolutionary lineages diverge. Any pair of lineages different enough to be called a phylum has been separate for a long time, and so has diverged a lot. So no confusion arises. But this, in turn, is because our phylum classifications are a snapshot of the present time. For all we know, two creatures of the same species right now will become the ancestors of different phyla in half a billion years' time. That does not affect our current classifications. (How could it, since we cannot see the future?)

But imagine travelling back along the line of ancestors from ourselves to the common ancestor we share with protostomes, the enormous group that includes arthropods, annelids, molluscs and many more. There must be such a common ancestor, whatever it was. Now imagine tracing (say) a fruit fly's ancestry back to the same common ancestor. A fruit fly and a human being clearly (we want to say) belong to different phyla. But at some point in tracing the two lineages, we will get to two creatures that are equally clearly of the same species. At some point in between, then, the lineages must have diverged far enough apart to be considered, at least, different species. Later they diverged enough to be considered different genera, and so on. But at what point, exactly, do two species become different enough to be different genera? At what point do two genera become different enough to be different families, and so on? Any answer we gave to this question would, once again, be drawing a line through a continuum. Dawkins makes this vivid with a thought experiment:

> It is all too easy, with hindsight, to think that because we recognise two ancient fossils as belonging to different modern phyla, those two fossils must have been as different from each other as modern representatives of the two phyla are. It is too easy to forget that the modern representatives have had half a billion years in which to diverge. There is no good reason to believe that a Cambrian taxonomist, blessedly free of 500 million years' worth of zoological knowledge, would have placed the two fossils in separate phyla. He might have placed them only in separate orders, notwithstanding the then-unknowable fact that their descendants were destined eventually to diverge so far as to warrant separate phylum status. (2004: 367)

The distinctions between major taxonomic groups, such as phyla, are not vague. It is precisely because they are highly divergent from each other that two groups are considered to deserve the name phyla. So we do not find ourselves embarrassed by intermediary cases. What is vague, however, is when, exactly, we should start calling a group a phylum rather than a mere order, when we should start calling it an order rather than a mere family, and so on.[3]

Yet much importance seems to be attached not just to what taxonomic group a creature belongs to, but to what rank a taxonomic group belongs to. Animals, we are told, make up a kingdom (Animalia); chordates make up a phylum (Chordata) within the animal kingdom. It is true that animals are a monophyletic group, and that chordates are a smaller monophyletic group within the animal kingdom. We could not say that animals were a phylum within the chordate kingdom. But why can we not say that chordates are an order within the animal phylum? The orders within the chordates would have to be demoted as well if we did this, but that would not be impossible. We even have classificatory terms such as "sub-order" for intermediate groups, so we would not have to demote at every level.

And it gets stranger. The tree of life does not look like Figure 8.1; rather, it looks like Figure 8.2 (although it has many more branches.) A lineage does not suddenly spring into existence as a phylum. Let us look at branches (1) and (2). We could, if we wished, call them separate phyla. If the divergence took place a long time ago – as is the case with all of the categories we actually do call phyla – then there is likely to be a high degree of disparity between those on one side of the divide and those on the other. But we could consider those creatures on the branch from which both (1) and (2) stem to belong to either phylum (1) or phylum (2). Thus, we could, if we wished, consider branch (1) to have diverged from branch (2), or vice versa. The way I have drawn it makes it look as though branch (1) diverged from branch (2), but I could have drawn it the other way around without changing any of the facts. Tetrapods – amphibians, reptiles, birds and mammals – are almost certainly descended from bony

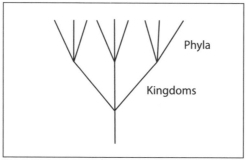

Figure 8.1 An unrealistic view of the "tree of life".

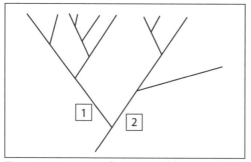

Figure 8.2 A more realistic view of the "tree of life".

fish. We generally treat amphibians, birds and mammals as classes. (Reptiles are a special case, as they are not in fact a monophyletic group.) But we treat present-day bony fish (*Osteichthyes*) as a class too, that is, as a clade of the same level as amphibians and the others. We could, if we wished, treat tetrapods as a mere subdivision of bony fish. The last common ancestor of tetrapods and bony fish was in all likelihood a bony fish by any reasonable standard, not a tetrapod. Mammals and birds are mere subdivisions of the tetrapods, so we would be justified in placing them not one level but two levels down in the hierarchy from bony fish. The reason we do not do this is, apparently, that we feel that the differences between mammals and the rest are of sufficient importance for mammals to merit the exalted rank of class.

What level we want to call a taxonomic group is, then, partly up to us. It is, however, constrained by the need to be consistent. The levels come in a certain rank order: kingdoms are divided into phyla, which in turn are divided into orders and so on. A phylum cannot be a subdivision of a class, and two creatures of the same class cannot be different phyla. But suppose we have already decided that group A is a class and belongs to phylum B. We might subsequently want to insert a further subdivision of B that includes all As but other things as well. We might solve this by demoting A to a class, or we might designate the new subdivision a sub-phylum.

The issue of whether a group deserves the name phylum is not settled by how big the group is. That is so whether we take into account number of species or number of individuals. The phylum Placozoa contains, so far as we know, just one species, but it is still considered a phylum for all that. A natural catastrophe could affect the organisms in one group more than others, but that would not justify us in demoting the group from a phylum to an order. Moreover, if a group is sufficiently homogenous to be considered a species, we consider it a species no matter how many of them there are. The world population of the common chimpanzee *Pan troglodytes* is currently estimated at 105,000, as compared to about six billion *Homo sapiens*, but both are considered species for all that.

The sole known representative of the phylum Placozoa is *Trichoplax adhaerens*, a nondescript multicellular animal. This means that, to the best of our knowledge, the phylum contains not just one species, but also just one genus, one family, one class and one order. It is, of course, possible that there was a more diverse range of Placozoa in the past, but, again, we need not be deterred from calling them a phylum if there was not. Why, then, is *Trichoplax adhaerens* given a phylum all to itself? It is because comparison of DNA tells us that it is sufficiently distantly related from all other creatures to belong to a different phylum from any of them. There are a number of different hypotheses about where exactly Placozoa belong on the tree of life (see Fig. 8.3).

Even if hypothesis (c) is true, Placozoa forms a group that is at least as "senior" in rank as Bilateria. But Bilateria includes chordates, arthropods, annelids, platyhelminthes (flatworms) and more (the name means having two sides, i.e. bilateral symmetry). Those groups are all phyla, so unless we wanted to demote all of them from this status, we have to accept Placozoa as at least a phylum as well. Indeed, we should probably consider it more senior than a phylum: a "superphylum" perhaps. And now look at Cnidaria on the tree: in all three hypotheses, Cnidaria is a branch that diverged earlier than Bilateria. But Cnidaria is considered a phylum, while Bilateria is considered more than a phylum. It looks as though something has gone wrong here. However, the potential disagreement is not about whether a group is a genuine clade, or about where it goes on the tree of life. The disagreement about where Placozoa

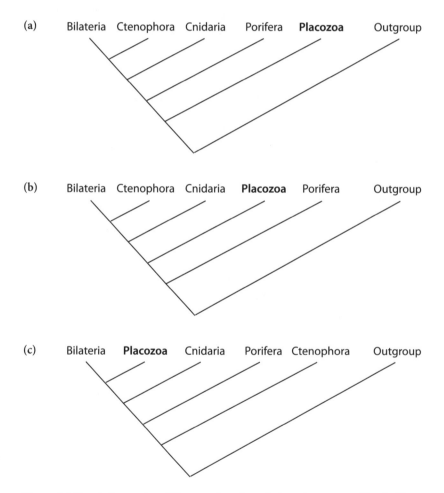

Figure 8.3 The phylogenetics of Placozoa, based on phylogenies presented by the University of California Museum of Paleontology (Collins n.d.)

go is because of insufficient evidence, not because different taxonomists have different preferences.

The element of choice regarding clade ranking can be illustrated by the way the hominid fossil record has been dealt with. Every now and then a newspaper bears the sensational headline "New Species of Human Discovered", or words to that effect. But it is equally common for some palaeontologists to dispute the claim that the new specimen should be considered a distinct species, and argue that it is a mere subspecies. Others might argue that it is a distinct genus. Those who like to demote groups – calling them subspecies rather than species, for example – are sometimes called "lumpers", while those who like to promote them are called "splitters". Species, as I shall show in Chapter 9, are often taken to be a special case, so the issue of lumping versus splitting is a little different if one of the possibilities under consideration is for a group to be considered a species. But when it comes to deciding whether a group is, say, a phylum or an order, lumping or splitting are really a matter of preference, constrained only by the requirement of consistency. Look again at Figures 8.1–8.3. Perhaps the most grievous example of splitting in taxonomy is the categorization of the *subgroups* of Bilateria as phyla, while Cnidaria and Placozoa, which ought to be of equal or higher ranking to Bilateria, are called phyla as well. (One could alternatively call the treatment of Cnidaria and the others a grievous example of lumping.)

The hierarchical terms phylum, kingdom, and so on may seem to be more trouble than they are worth. Lumpers and splitters need not be disputing any matter of fact whatsoever. The designers of the highly ambitious web project Tree of Life (www.tolweb.org/tree) completely sidestep the issue by simply referring to "containing groups". This may be the wisest approach. Rather than agonizing over whether some group is a phylum or not, we only need to know which branch it is on the tree of life. It would be no harm if the categories kingdom, phylum, order and the rest were abandoned. That does not mean we should abandon the *names* of the different kingdoms, phyla and so on. The statement that such-and-such a creature is a chordate carries genuine information. But the statement that Chordata is a phylum does not. The term "species" has been treated rather differently, however, as we shall see in Chapter 9.

9. Species and their special problems

As I mentioned in Chapter 8, the creators of the admirable Tree of Life internet project have not bothered with ranking terms such as phylum, class and so on. That, in my view, is one of the reasons why it is admirable. However, they do retain the ranking term "species". Why do biologists who think that it does not matter whether a group is a phylum or not think it does matter whether a group is a species or not?

The word "species" has a historic resonance. Entities of all kinds fall into nested categories: gold is a metal and a chemical element; "3" is a natural number, an integer, a rational number and a real number. (Admittedly, the chemical tree is nowhere near as profuse as the tree of life; there are only a hundred-odd elements after all.) It would be correct to say that you are a human, a primate, a mammal, a tetrapod, a chordate, a deuterostome, an animal, a eukaryote and a living thing. (I have left out some categories in between.) But, by longstanding tradition, when we say what species something belongs to, we are saying what it *is*. Aristotelian essences are connected differently to species than to other levels of classification. For Aristotle, to say what species something is, is to give its essence: "essence will only belong to species of a genus, and to nothing else" (*Metaphysics* Z, 1030a13). To say that I am an animal is to give too little information; to say that I am an Irishman is to give too much information, to give my essence. My essence is that I am human. (For Aristotle, gold is also a species: a species of the genus metal.) Species, then, are a special kind of kind.

Biology and meta-biological thinking have changed considerably since Aristotle's time, for better or for worse. Nonetheless, the idea that there is something special about the rank of species has kept a grip. As we saw in Chapter 1, Darwin tried to break this grip, emphasizing many times in the *Origin* that one person's species is another's mere variety. But despite the veneration in which Darwin has long been held, it was standard practice well into the twentieth century to take a "typical" member of a species as representing the entire species. Moreover, we talk of "endangered species", and rates of extinction are often measured in number of species going extinct in some unit of time. The feeling that the species ranking is special is not adequately explained as a leftover of Aristotelian ways of thinking. Nor is it to be explained by the fact that some of our longstanding vernacular classifications of animals are at the species level: dog, human being, horse and so on. In any event, this is only true of a very small range of species; the vernacular classifications "monkey" and "spider" pick out larger groups. Whether a group is a species is generally felt to have a *scientific* significance that other rankings do not have. Species, it is often said, are what evolve.

One reason this is held to be the case is that genetic changes can be brought about by new combinations of genes through sexual reproduction. Clearly, this can happen between any two human beings of the opposite sex, or their off-spring if they are the same sex, whereas it cannot happen between human and non-human beings. The other producer of genetic change, mutation, leads to a similar moral. A new gene can spread to other members of the same species, but not to members of a different species. Moreover, the members of a species are likely to have roughly the same physical makeup and roughly the same way of making a living. This means that a change to the environment is likely to challenge all the members of the species in the same way. Not only does it put them all under selection pressure, but it puts them all under the same selection pressure. It presents them all, that is, with the same problem. It is true that a more local change – one that only occurs in one small part of the species' geo-graphical range – will only affect some of the species, so that only those will have to adapt or die. But those local selection pressures are precisely one of the ways in which new species come about. We shall see later that things are not in fact as simple as I have described. But for the moment let us accept that there is a *prima facie* case that the species ranking is important for evolution in a way that the other rankings are not.

9.1 The interbreeding criterion

Like phyla and classes, species are monophyletic groups. The species ranking has its place in the hierarchy in that a species is a junior rank to a genus, a genus to a family and so on. But in saying that two groups are distinct species, we are saying more than just that they are distinct monophyletic groups with less divergence than two genera. There is a *further criterion* that is usually used to decide if a group is a species. A common way to deal with this is to say that two individuals are the same species if they can interbreed with each other (or, if they are the same sex, at least if their opposite-sex parents could have interbred with each other). This is called the *biological species concept*, and has its origin in the work of Ernst Mayr (1942). This is why dogs are considered a single species despite the fact that they come in a wide variety of shapes and sizes. There are divisions of species elsewhere in the animal kingdom between creatures that look much more alike than an Irish wolfhound and a Chihuahua: common chimpanzees and bonobos, for example. A biologist who did not already know, and did not have access to genetic data, might well look at our various breeds of dogs and decide that there were several different species. But the criterion of interbreeding appears to settle the question. Calling a group a species, then, is not as subject to preference as calling a group a phylum. Unfortunately, things are not quite as clean-cut as this suggests, for several reasons.

9.1.1 Can they actually interbreed?

First, consider the question of whether an Irish wolfhound and a Chihuahua could *actually* interbreed. They certainly could not without artificial assistance. A male Chihuahua could only impregnate a female wolfhound by artificial insemination, and a female Chihuahua could not carry puppies as large as the hybrids would probably be in her womb. Nonetheless, we can say that *in principle* they could interbreed. This can be supported by a more robust criterion, that of genetic compatibility. A dog that had half Chihuahua genes and half wolfhound genes would be perfectly viable, and moreover would be able to produce offspring itself. This is in contrast to a mule, which is sterile, and so we say that horses and donkeys cannot interbreed in the proper sense of the word. In actuality, groups are sometimes considered distinct species even though they can in principle interbreed, that is, even though they are genetically compatible in the sense just explained. Dogs and wolves can interbreed, but they are considered different species. This happens, for example, in cases where two groups cannot actually interbreed. Sometimes this is because of logistic considerations similar to the Chihuahua/wolfhound case, but sometimes it is simply because of geographical separation. This last reason seems clearly insufficient to consider two groups to be distinct species. Would we consider Native Americans before 1492 to be a separate species from Europeans?

But the reason why the interbreeding criterion is taken to be important is that it allows genes to spread around the group, thus making it the group that evolves. This suggests that *actual* ability to interbreed is required. Moreover, if the two creatures in question are geographically separated *and* morphologically very different, then we cannot say that an environmental change will affect them in the same way. Such considerations have led some theorists to propose an extreme "splitter" approach (e.g. Rosen 1978). At the other extreme are extreme "lumpers", who argue that only groups that are completely unable to interbreed should be considered different species (e.g. Grant 1981). The super-splitter approach seems highly impracticable. It is always possible that some, even very small, geographical variation, or morphological difference, will mean that *some* environmental change will not affect the whole group in the same way. Moreover, it seems clear enough that Native Americans and Europeans *were* the same species before 1492. Suppose, then, that we accept the genetic compatibility criterion as sufficient for single-specieshood, and accordingly rule that species classifications that fail to respect this are just *wrong*, and that the perpetrators should be reprimanded for indulging in unwarranted splitting. Would we then have a clear answer to whether any group is a species?

9.1.2 Ring species

Unfortunately, we would not have a clear answer to whether any group is a species. A further problem arises with what are known as *ring species*. Dawkins (2004: 252–61) drives this point home with the example of the salamanders of the western USA. In southern California there are two distinct species of salamander by the criterion of genetic compatibility: they cannot interbreed and that is that. So far so good. But if we take a group of one species – call it *A* – and compare it with its group of neighbours a little further north – call it *B* – we find that they can interbreed. Go a little further north again, and we find another group, *C*, which can interbreed with *B*, and so on around the ring, until we get to (let us say) group *F*, which cannot interbreed with *A*. So we have: *A* can interbreed with *B*, *B* with *C*, *C* with *D*, *D* with *E* and *E* with *F*. But *F* cannot interbreed with *A*. So, which species do we put the intermediate groups into? *A*, or *F*, or some other species? By the genetic criterion we have good reason to consider *A* the same species as *B*, *B* the same species as *C* and so on. But *by the same criterion* we have good reason to consider *A* and *F* different species. So the logical outcome seems to be that *A* and *F* both belong and do not belong to the same species. But this is a contradiction, so something seems to have gone wrong somewhere.[1]

What it suggests is that we should classify creatures that are *extremely unlikely* to interbreed as separate species, rather than requiring that they be completely unable to. This would mean that whether a group is a species is a matter of degree, as Darwin had said. That is, sometimes it is clear that two groups are different species, but not always.

9.1.3 Asexually reproducing organisms

But there is still more bad news. Ring species, it might be suggested, are not all that common, so perhaps we should not abandon a criterion that works well in most cases. Maybe so, but unfortunately asexually reproducing organisms are *extremely* common. Asexually reproducing single-celled organisms form by far the majority of living things. There are far more individual single-celled organisms than anything else, and just in case you think it is a bit unfair to count one single-celled organism as equal to one multi-celled organism, the *actual mass of living matter* (biomass) composed of single-celled organisms is believed by many biologists to be greater than the rest as well. There are also multi-celled organisms that reproduce asexually, such as the already mentioned rotifers. But when organisms reproduce asexually, there is no sense in which they can be said to interbreed, either actually or in principle. But how, then, can we speak of groups of these organisms as constituting species, as we do? There is still genetic similarity, of course, to guide our decisions. But within the asexually reproducing world there is a continuous spectrum running from clone groups that are genetically identical, to differences as great as those that separate whole kingdoms in the sexually reproducing world. Clearly, being a clone group is too strong a requirement for single-specieshood. Imagine if we tried to apply it to human beings: only identical twins and the clones that science-fiction writers promise us are coming would qualify. We seem to now have a situation where deciding if a group is a species is a matter of preference. In the case of ring species, we might say that the genetic compatibility criterion just happens not to work. At least we know what it would mean for it to work. But in the case of asexually reproducing organisms, it makes no sense at all.

Similar problems arise with many plants, which can form hybrids much more readily than animals. In many cases those hybrids, unlike mules, can breed. Once again, the criterion of genetic compatibility just does not work. But, it might be said, the fact that the genetic compatibility criterion does not work for very many, even the vast majority, of living things, does not imply that it never works. Perhaps the concept of species should only be applied to sexually reproducing organisms that do not form ring species and that do not produce fertile hybrids. As a further alternative, a *pluralistic* approach to deciding if some group is a species or not has been suggested, where the word "species" has a different meaning in different parts of the living world. I shall return to this point in §9.3. For the moment let us assume that we can live with at least one of these possibilities. Surely, it might be suggested, a criterion that works for some cases is better than no criterion at all? At least you can be sure that you are the same species as a person on the other side of the world, and that you are both a different species from a chimpanzee.

9.1.4 Vagueness over time

But the problems do not stop there. You can be sure that you are the same species as Alfred the Great as well. We only need to stretch a little the sense in which you can *in principle* interbreed to show this. If you had a time machine, or if Alfred's DNA had somehow been preserved, then you (or, if you are male, your female parent) could produce a fertile offspring with him. So far, so good. Go further back in time, however, and things are not so good. It is beyond dispute that the people of one generation are the same species as those of the generation before it. Common sense and the genetic compatibility criterion both say the same thing. But trace your ancestry back a few hundred thousand generations, and you will eventually get to creatures that are indisputably a different species from you. They neither look like *Homo sapiens*, nor are they genetically compatible with us. But every successive generation in that journey is the same species as the one before. There was no point at which a non-*Homo sapiens* gave birth to a *Homo sapiens*. We are faced with exactly the same paradox as with ring species. It can be shown by perfectly logical reasoning that the intermediaries are both the same species as us, and the same species as something that is not the same species as us.

But, unlike the case of ring species, this paradox arises not just in a few unusual cases, but throughout the living world. As we saw in Chapter 1, it is a central pillar of Darwin's theory of evolution that new species arise by the accumulation of tiny steps. And, as we saw in Chapter 4, the reasonable version of punctuated equilibrium does not challenge that. So, even if we restrict it to sexually reproducing organisms that are not ring species, and do not produce fertile hybrids, the interbreeding criterion for single-specieshood does not work over time. Just as no answer can be given to the question "Are the salamanders of the western USA one species or two?", it appears that no answer can be given to the question "When did such-and-such a species come into existence?" It would be like asking: on what day did John stop being an adolescent and start being an adult? For legal purposes such as being allowed to vote or buy alcohol, we do provide exact answers to this question. But we do this because we have to, not because we really think that adulthood is such a precise concept that it begins on a particular day. We would perhaps be comfortable with giving an *approximate* date to when John became an adult: perhaps when his behaviour started to show sufficient signs of maturity. Palaeontologists do, when they have enough information, give *approximate* dates for the appearance of new species, and they do not appear to be talking nonsense when they do so. Several species of dinosaur really did arise and die out long before *Homo sapiens* arrived. Despite the paradoxes that arise when we allow our categories to be vague, biologists are perfectly right to allow them to be vague. And despite the paradoxes, biologists are not reduced to talking nonsense.

9.2 Species as individuals

Throughout this chapter so far, and Chapter 8, I have been writing about biological *kinds*. Gold is a kind, and I have been writing as if horse, dog, mammal and chordate are all kinds as well. One thing that has emerged in the foregoing discussion is that the criteria for including something in a biological kind are very different from the criteria for including them in a chemical kind. Something is included in a chemical kind in virtue of a characteristic it *currently* has. Pheneticists such as Sneath and Sokal propose that we take a similar approach to classifying living things, but this approach seems to be beset with problems. Moreover, chemical kinds are free from vagueness – something is either gold or it is not – whereas biological kinds are vague, and seem to be inescapably so. This is the case whether we take a phenetic or a cladistic approach. One moral we might draw from this is that when we use the term "natural kind" we mean different things in different cases: that is, the sense in which a biological species is a natural kind is not the sense in which a chemical element is a natural kind. This moral is drawn by David Hull:

> From the beginning, a satisfactory explication of the notion of a natural kind has eluded philosophers. One explanation for this failure is that the traditional examples of natural kinds were a mixed lot. The three commonest examples of natural kinds have been geometric figures, biological species, and the physical elements. By now it should be clear that all three are very different sorts of things. No wonder a general analysis, applicable equally to them all, has eluded us.   ([1978] 1994: 208)

Hull proposes a radical solution to this problem, which was also proposed by Michael Ghiselin (1974). Rather than consider species to be kinds at all, we should consider them to be *individuals*.

At first sight, this seems like a very strange suggestion. Surely, you might think, horses (say) constitute a kind, whereas Black Beauty is an individual horse? However, Ghiselin and Hull are not saying that Black Beauty is not an individual. They are saying that the same reasons we consider Black Beauty an individual justify calling the species *Equus caballus* an individual as well. There seems to be no problem in an individual being part of a bigger individual. You have *individual* cells in your body, which are part of you, but you are no less an individual.

What are the grounds for calling a species an individual, and not a kind? Recall that the criteria for calling something gold appeal to properties that that thing currently possesses. If it has atomic number 79, then it is gold. This means that if something with atomic number 79 were to appear at any time, anywhere in the universe, it would be gold. If all the matter of atomic number 79 were to disappear, there would be no gold. If any were to subsequently appear, it would be gold. Moreover, if scientists synthesized atoms of atomic number 79 in a laboratory, they would be gold atoms. Thus, the kinds in chemistry are *spatiotemporally unbounded*. This is a straightforward consequence of the fact that whether something is gold or not depends on whether or not it *currently* possess

149

such-and-such a property. By contrast, on the cladistic conception of taxonomy, which is the currently orthodox view, it is a necessary condition for something being a member of the species *Equus caballus* that it share a common ancestor with all and only members of that species. Thus, no matter what properties a creature possesses – no matter how much it looks like a horse, walks like a horse, or neighs like a horse – if it does not have the ancestry common to horses, it is not a horse. This is a straightforward consequence of accepting that being a monophyletic group is a necessary condition for being a species, even if it is not a sufficient one. Thus, if the horse species goes extinct, but creatures that are exactly the same as them come into existence later on, those creatures will be not be horses. Species, unlike chemical elements, are *spatiotemporally bounded*.

Ghiselin and Hull argue that this spatiotemporal boundedness makes species more like the things that are undoubtedly individuals, than like the things that are undoubtedly kinds. A person is an individual. But people go through many changes in their lives. It is perfectly conceivable that there is a five-year-old somewhere who is more like you when you were five than the present-day you is like you when you were five. But this would not for a moment lead you to doubt that you *are* the same individual as the five-year-old you, and that that other person is *not* the same individual as the five-year-old you. Similarly, if a scientist were to make an exact copy of you, it would not be you. That would be the case even if the copy were a genetically exact replica, for we all know full well that a pair of identical twins is two people, not one. (Much science fiction plays with the idea of cloning, and exploits a fear that being cloned would somehow blur the boundaries of individuality. The authors, or their audiences, seem to have forgotten what we all know about identical twins.) In this respect, then, species seem to "behave" more like individuals than like kinds.

Things are perhaps not as simple as this suggests, however. Imagine a *Jurassic Park*-type scenario. Assuming for the sake of argument that the logistical problems are as easily surmountable as Michael Crichton glibly assumes that they are, would we call these creatures *Tyrannosaurus rex* and so on? If we do, are we not accepting that, contrary to what I said above, a species could go extinct and then come back into existence again? Even with the arguments I have just summarized, there is still some reason to say: yes, they are members of the species *Tyrannosaurus rex*. The hypothetical scenario in Crichton's novels is that the new creatures were produced using DNA from actual, undoubted tyrannosaurs. So there is a causal connection between the undoubted tyrannosaurs and the new creatures, albeit a more tenuous causal connection than in the normal case. Similarly, if your sperm or eggs were frozen and used to produce a baby after you were dead, we would still probably say that you were the baby's father or mother. By contrast, if we synthesized gold atoms out of atoms of some other element, it would surely *not make any difference* whether those atoms had in turn been synthesized out of gold atoms or not. Either way, it would still be gold.

There does not have to be any causal connection whatsoever between the new atoms and any previously existing gold atoms.

Let us press this issue a little further. Suppose that, instead of using DNA from tyrannosaurs, the scientists synthesized DNA of the right kind using the raw materials cytosine, guanine and so forth. Would we *now* say the new creatures were tyrannosaurs? Maybe; maybe not. On the one hand, the causal connection that was there on Crichton's scenario is no longer there. On the other, the scientists have *copied* tyrannosaur DNA, so there is still *a* causal connection, albeit one that is even more tenuous. The causal connection goes via the knowledge and the intentions of the scientists. For the causal connection to be totally severed, we would need a situation where new creatures exactly like tyrannosaurs came about, which were neither born of undoubted tyrannosaurs, not made by scientists using their knowledge of the tyrannosaur genome. Perhaps they could evolve by some process of massive convergence and/or coincidence. Or perhaps scientists, without intending to, could produce DNA that was just like tyrannosaur DNA. Note, however, that these complicated questions do not arise with gold. The criterion for being gold can be applied to new atoms that are formed totally by chance, or by natural forces without any human intervention whatsoever. So perhaps the very fact that these complicated questions arise gives further support to the view that species are not kinds.

Let us return to the cloning issue, however. I deliberately avoided saying *how* the scientist makes this exact copy of you. On the usual understanding of "cloning", it is done by taking some DNA from you, just as in *Jurassic Park*. We would, for that reason, have no problem saying that the new person produced was the same species as you. But the general consensus (certain science-fiction writers apart) is that the new person is not *you*. There is, once again, a whole host of possible variations on this theme. Suppose an atom-by-atom replica is produced while, at exactly the same time, you are annihilated. Is that new person you? Many people would probably say yes, and at least the example would give us pause in a way that the cloning example does not (or at least should not) give us pause. But if I am right that in the cloning case the new person is clearly not me, then the requirement for being the same individual is much stronger than mere causal connection *of some kind*. In fact, the causal connection between you and a clone of you is of exactly the same kind as that between the original tyrannosaurs and the new creatures. Perhaps some other scenario where the causal connection was weaker than in the Crichton story would lead us to say that the new creatures were not tyrannosaurs. And it is likely that in both the individual person case and the species case there are borderline scenarios where we would find it difficult to decide either way. Gheselin and Hull have made a good case that species do not behave exactly like the things that are undoubtedly kinds. But, I suggest, perhaps they do not behave exactly like the things that are undoubtedly individuals either.

9.3 A pluralistic approach

Brent Mishler and Robert Brandon (1987) have argued that those who hold the "species as individuals" view are relying on an oversimplified conception of what an individual is. There are, Mishler and Brandon argue ([1987] 1998: 302–3), at least four different reasonable criteria for whether something is an individual, and these have been conflated by the likes of Gheselin and Hull:

(i) *Spatial boundedness* – the fact of being confined to one part of the world. A "proper" kind, such as gold, may appear anywhere in the world, and is still gold no matter where it appears. This does not appear to be the case with species, however.

(ii) *Temporal boundedness* – the fact that a thing has a single time when it begins, and, if it has an end at all, a single time when it ends. That is, when it ends, it does not come into existence again, even if something else exactly like it does.

(iii) *Integration* – the fact that the parts of something interact with one another. Mishler and Brandon ask, "does the presence and activity of one part of an entity matter to another part?" (*ibid.*: 302). For example, the activities of the heart and the brain matter to each other. If one malfunctions, it is liable to affect the other. Groups of organisms that can actually, rather than just in principle, interbreed (see §9.1.1) are integrated in this sense. A catastrophe that affected one part of the group would make a difference to the rest, as it would reduce their gene pool.

(iv) *Cohesion* – the fact of responding *as a unit* to events. As an example of this, Mishler and Brandon offer the response of a corporation to a crash in the stock market. Presumably, the response of a person to a bullet in the head would also be an example.

Criteria (iii) and (iv) are vague. Clearly, not every activity of every part of some-thing has to matter to every other part for it to be considered an individual. A cut on your toe does not affect the rest of you. But at least some things that happen to your toe affect the rest of you: your ability to walk would be affected by a severe injury to it, for example, and that in turn could affect your ability to gather food and escape from danger, and your general level of happiness. And at least some things that go on in other parts of your body affect your toes. Conversely, the presence and activity of plenty of other things in the world make a difference to you without those things being considered part of you. One could still make a case, however, that there is a non-trivial difference of degree here. You are by now, I take it, getting used to biological concepts in general being vague. So we can hardly hold this against criterion (iii). As for criterion (iv), what Mishler and Brandon call cohesion, a similar story can be told. An earthquake could throw you and the entire contents of your apartment into the air as one, so it could be said that you and your worldly goods are responding to that event in unison. But you would not consider yourself and the contents of your apartment to be one individual. Once again, however, there is clearly a difference of degree.

In any event, it is not the vagueness of the concept of individuality that concerns Mishler and Brandon. Rather, it is the fact that the different criteria pull different ways in some cases. For example, a group that was actually able to interbreed but was spread over a wide area would probably not respond as a unit to a change in climate. Imagine, for example, that in the area there are differences in altitude, or differences in vegetation. The change in climate, even if it affected the whole area, would probably affect different parts of it in different ways. The creatures in one area might find themselves facing a famine, while those in another are unaffected. Nonetheless, the group as a whole could still fulfil at least the first two criteria for being an individual.

One might object that it is really the first two criteria that count. The consensus view on taxonomy is cladistic, that is, sees descent as the deciding factor. The "species as individuals" view of Gheselin and Hull does not go against this. But integration and cohesion are *current* properties of a population, so to consider them as criteria for membership in the individual-that-is-the-species is surely a throwback to the phenetic approach to taxonomy. Hence, it might be concluded, only criteria (i) and (ii) should count.

However, things are not so simple. Recall that I said that being a monophyletic group is a *necessary* condition for being a taxon, not a sufficient one. If a group changes sufficiently greatly, then surely it is reasonable to consider it a new species. If you do not find this plausible, imagine a *really big* change. There must come some point at which even the most stubborn lumper will say it is a new species at least. If being a monophyletic group were the sole and sufficient criterion, we might as well consider the whole of life a single species.

In any event, Mishler and Brandon argue that criteria (i) and (ii) can pull apart as well. The example they offer is where a hybrid of two plant species arises, and goes on to reproduce, being distinctive enough and stable enough to be considered a species. (This is actually quite common in the plant kingdom, and I shall discuss it further shortly.) Their point is that the same two species could hybridize several times. Thus the hybrid species, although spatially bounded, would not be temporally bounded; it would have more than one origin.

Elsewhere, Mishler and Michael Donoghue ([1982] 1994) argue for *species pluralism*. That is, they argue that a criterion for specieshood that works in one part of the tree of life may not work for other parts. I noted above that the Linnean system of classification worked reasonably well for those parts of the animal kingdom that are nearest to us, and less well for parts further away. Mishler and Donoghue suggest that, similarly, biologists have tended to favour approaches to classification that work well for the parts of the tree that they know best:

> It seems clear that the group of organisms on which one specialises strongly influences the view of "species" that one develops. It also seems clear that in order to fully appreciate biological diversity (for purposes of developing general concepts), it is essential to study a variety of different kinds of organisms, or at least take seriously those who have.
>
> (*Ibid.*: 219)

As we have seen, the commonly cited criterion for specieshood – the biological species concept, which is in terms of capacity to interbreed – seems to be inapplicable to asexually reproducing organisms. Because of the relative commonness of hybridization in the plant kingdom, it seems inapplicable there also. Awareness of problems such as this have led some people to develop alternative criteria for specieshood.[2] Essentially, these alternative approaches echo the different approaches to taxonomy in general that we looked at in Chapter 8. There are approaches based on character (like phenetics), approaches based on descent (like cladistics), and approaches based on combinations of the two (like the evolutionary approach). As regards a character-based approach, the same problems arise as arise in connection with phenetic classification in general. It is not capable of yielding an unambiguous answer in all cases.

A cladistic approach may seem more promising, but it faces a new problem of its own. It would fall down in the plant hybrid cases. Consider how we would graphically represent a hybridization on a tree of life diagram. We would have to show two branches growing together to form one branch. Since we are talking about *fertile* hybrids, we would also have to show that branch growing upwards, and perhaps dividing into smaller branches as well. But what happens when we try to apply the concept of a monophyletic group here? The new hybrid and all its descendants form a monophyletic group, sure enough. But monophyletic groups form parts of bigger monophyletic groups, so we should expect our new one to do so. The plants that hybridized to form our new type may have been parts of well-behaved monophyletic groups. But, since *ex hypothesi* they were of different species, they would have to have been of different monophyletic groups. But, does our new hybrid belong to the same monophyletic group as its mother or its father? Its mother, the rest of her species and all their descendants form a monophyletic group that includes the new hybrid and all *its* descendants. But its father, the rest of *his* species and all of *their* descendants form a monophyletic group that includes the new hybrid and all of its descendants as well. So our new hybrid belongs to two distinct but overlapping monophyletic groups. We expect any given organism to belong to just one species, just one genus and so on. Although the boundaries of species are fuzzy, and although I have urged that we should not bother about whether a group is a phylum or an order, I have been assuming that the concept of a monophyletic group is itself unproblematic. Now we find that it leads to yet more paradoxes. Moreover, in cases where two plant species hybridize, they often do so *more than once*. The products of these different hybridization events can often interbreed, and they can evolve like any species. This may lead to a situation where the hybrids, as a single interbreeding group, end up unable to interbreed with the ancestral species, but able to interbreed with each other. In this event, the products of the different hybridization events would be a single species by the interbreeding criterion, but two different species by the monophyletic criterion (see Fig. 9.1). Moreover, the two hybrids are also likely to resemble each other enough to be one group by any reasonable phenetic criterion. As we saw in Chapter 8, combining phenetic and cladistic approaches does not solve the problem, as we would have the problem of relative weighting. The same would be true if we added the interbreeding criterion to the mix. It seems, then, that we just have to use different criteria on different occasions. The case for pluralism seems to be a strong one.

In fact, a case could be made for a *higher-order* pluralism. Just as Mishler and Donoghue argue that we should use different criteria for specieshood in different parts of the living world, so too, perhaps, we should sometimes use different criteria for deciding if something is an individual or a kind. Gheselin and Hull argue that the species-concept behaves more like individual-concepts than like kind-concepts. But Mishler and Brandon, as we saw, argue that different criteria for whether something is an individual sometimes pull in different directions. What I have suggested is that, in some ways, species do not behave like individuals. The causal continuity, it seems, can be much more tenuous for a species than for an undoubted individual such as a person. Perhaps, then, we should simply conclude that species are in some ways like individuals and in

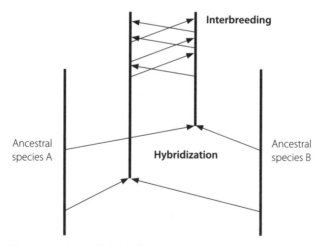

Figure 9.1 Do the hybrids form one species or two?

some ways like kinds. Or perhaps we should say that species are kinds, but not the same kind of kinds as other kinds. It is not clear from philosophical argument or scientific practice whether species are individuals or kinds. So perhaps the fact that it just *seems more natural* to call them kinds should be given some weight. They are, perhaps, *historical* kinds, unlike chemical elements.

Recall that Hull said that philosophers' stock examples of kinds – geometrical shapes, biological species and chemical elements – are "a mixed lot". This seems to be true, but that does not by itself give us a reason to conclude that not all of these are kinds. Still less does it give us reason to conclude that biological species should be taken off the list. If we can say that species is not a unitary concept, then perhaps we can say that *kind* is not a unitary concept either.

Considerations of taxonomy have led us to a number of paradoxes. Some of these, I have suggested, can be avoided by abandoning taxonomic practices that are more trouble than they are worth; for example, the practice of ranking clades as kingdoms, phyla and so on seems to serve no purpose. Both the vagueness and the apparent need for pluralism indicate that biological concepts do not have the pristine precision that we expect from scientific concepts (such as chemical elements). Yet biology seems to be a science in perfectly good working order. Philosophers of science have traditionally focused their attention on the physical sciences. Yet it is open to question just how applicable their philosophies are to biology. I shall take this up in Chapter 10.

10. Biology and philosophy of science

As I pointed out in the Introduction, philosophers of science have often concentrated their attention on the physical sciences. Often, indeed, the physical sciences have been taken to be examples of what a good science should be. For example, Popper illustrated his falsificationist criterion for separating the scientific sheep from the pseudo-scientific goats by taking Einstein's theory of relativity as a clear example of a sheep. One feature that is often held to distinguish the physical sciences is their possession of strict, exceptionless, mathematical laws (or, as some prefer to call them, simply "laws"). The speed of light is a strict mathematical constant, and the atomic numbers of chemical elements are exact numerical values. Sciences that lack such laws, such as meteorology or economics, are often relegated to the inferior status of "special sciences". The main focus of this chapter will be on the features of biology that make it different from the allegedly exemplary physical sciences. Recall my earlier quotation from Dawkins to the effect that biology is a more unified science than physics. I shall argue that, when we examine what the unifying feature actually is, we can see that biology is more than a "special science". In some ways it is exemplary.

10.1 Lawlessness in biology

There is reason to think that, like the special sciences mentioned above, biology does not have strict mathematical laws of its own. There are, as in any science, generalizations. But these generalizations have a habit of proving to be: (i) not distinctive to biology; (ii) not strict, exceptionless, mathematical laws; or (iii) not laws at all. That is, put in more positive terms, the generalizations found in biology are: (i) laws that belong to other sciences; (ii) *ceteris paribus* laws; or (iii) true by definition.

10.1.1 Laws belonging to other sciences

There is no doubt that laws are employed in explanations of biological phenomena. The processes that go on at the sub-cellular level, including photosynthesis, DNA-transcription, passage of substances in and out through cell membranes and so on are explained and understood by means of strict laws. But those laws are laws that are not distinctive of biology. Biochemistry is so called because it is a branch of chemistry. Whether biochemists belong to departments of biology or chemistry in universities, the laws they use in their explanations are chemical laws. That is, they are the laws that govern the combinations and other behaviours of the chemical elements of which living tissues are made. Artificially synthesized compounds with the same chemical makeup as the ones that occur naturally in animals and plants, would behave in exactly the same way as those natural ones. We might, however, be reluctant to call the artificially synthesized ones "biological". In any event, the laws governing both natural and artificial versions of biochemical substances, are derived, and flow smoothly from, more basic laws governing how chemical elements in general combine, that is, the laws based on the valencies of the various elements. For example, lithium, sodium and potassium all combine with other elements in the same way because they have the same valency. These laws clearly belong to chemistry, and are not distinctive of biology.

There are also physical laws involved in the proximal explanation of how the various parts of organisms work. Laws explain how the heart pumps blood, how birds fly and so on. But those laws are general laws of hydraulics and aerodynamics respectively. That is, they are laws of physics. The laws of hydraulics that explain how the heart pumps blood also explain how man-made pumps work. The laws of aerodynamics that explain how birds fly also explain how aeroplanes fly. This is not altered by the fact that birds and aeroplanes do not work in the same way. The aerodynamic principles of lift, drag and thrust are the same in both cases. Differences between the flying capacities of birds and those of aeroplanes are explained by *physical* differences between them. The fact that birds can fly by flapping their wings and aeroplanes cannot is explained by the fact that birds are lighter than aeroplanes. The fact that aeroplanes can go faster than birds is explained by the fact that aeroplanes have engines. That is to say, the differences are not explained by the fact that birds are biological and aeroplanes are not. The explanation of how birds fly is in terms of laws that are common to the physical world in general, not laws that are distinctive of biology. This holds for proximal explanations of the operations of organisms' parts in general.

10.1.2 Ceteris paribus *laws*

There are, however, some instances that we might want to call laws that are distinctive of biology. One example I mentioned in Chapter 5 is Williston's law: in any lineage where there are serially homologous parts, the number of those parts tends to decrease, while the diversity and specialization of different variations tends to increase. A second example is a mathematical law relating brain mass to overall body mass in mammals (see Martin 1981; Dawkins 2004: 69–76). This law holds good over six orders of magnitude, that is, from mouse-sized to elephant-sized. Admittedly, it has a margin of error, but that is not the same as having exceptions.

Unfortunately, however, these laws do have exceptions. The most obvious exception to the brain-size law is (you probably guessed) human beings. Our brains are bigger, given our body mass, than this law predicts. Our nearest relations' brains are too, but not to the same degree. And (you probably also guessed) the brains of dolphins are almost exactly as "overlarge" as ours. The discovery of a discrepancy like this may prompt a search for a special explanation for why the law has been broken in these cases. Indeed, there has been much speculation, all inconclusive, about why human beings' brains became so big. Likewise, there are exceptions to Williston's law. But such exceptions do not cause as big a stir as the discovery of an exception in a law of physics would. The discovery of an anomaly like this in physics, where an exception was found to a previously well-established law, would sooner or later cause a crisis in physics. (Popper and Kuhn agreed on this; they just disagreed as to whether it was sooner – Popper – or later – Kuhn.) These distinctively biological laws, by contrast, are tolerant of exceptions.

This kind of law is called a *ceteris paribus* law, from the Latin for "all things being equal". In some ways it is an unfortunate term, but we seem to be stuck with it. Perhaps "rule of thumb" would capture the idea better. Many philosophers of science think that the reason *ceteris paribus* laws do not lose their status when exceptions come along is that they have a fairly low status anyway. Just as a mere commoner can suffer indignities that the monarch cannot, mere *ceteris paribus* laws can suffer the indignity of having exceptions, but strict laws cannot. Some philosophers of science go so far as to say that *ceteris paribus* laws do not deserve the name of laws at all (e.g. Woodward 2002). Yet, whether we call them laws or not, their role in many sciences is not negligible. They allow non-trivial predictions to be made, and not just in biology. Although we often feel otherwise, the Met Office is more likely to get its weather predictions right than are you or I. Economists working on the basis of *ceteris paribus* laws such as "if the government raises interest rates, inflation will go down" are able to predict economic trends, if not as accurately as we would like, still more accurately than the person on the street. Moreover, the exceptions, when they occur, can often lead to fruitful lines of enquiry into why the laws were broken, even though it would not be a disaster should these enquiries prove unfruitful. The search for an answer to why human brains (and dolphin brains) got so big may yield an answer some day. It does not seem, then, that *ceteris paribus* laws deserve the disdain that they receive in some philosophical quarters.

Nonetheless, it might be argued, the disdainful ones may have a point. It is plausible to suggest that, even if *ceteris paribus* laws are better than nothing, strict (or "real") laws would be better still. Strict laws, perhaps, are something that every science ought to aspire to, *ceteris paribus* laws being something that they make do with in the meantime.

10.1.3 Truths by definition

There are, however, some generalizations in biology that do not tolerate exceptions, and that are distinctively biological. Recall that I said in Chapter 8 that the monophyletic criterion for counting a group as a genuine biological taxon works well for at least a great many cases. If we take one of these cases, then, we can say, "the superorder Salientia [frogs and toads] is a monophyletic group". This leads to the following universal generalization: "If x is a salientiate, then x shares a common ancestor with all and only salientiates". This looks like a law in the full, exceptionless sense. But, although it might be exceptionless, perhaps it is not best thought of as a *law*, strictly speaking. Perhaps it is better thought of as *true by definition*, if it is true at all. That is, perhaps it cannot be counted as an empirical fact in the way that "human beings' closest relatives are chimpanzees" can. But the laws of physics state mathematical relations that are not simply true by definition. It is not simply true by definition that the speed of light in a vacuum is 299,792,458 metres per second (denoted by c). This is because we can identify something as light independently of knowing what its speed is. However, we could not identify some group as belonging to Salientia without also identifying it as sharing a common ancestor with all and only salientiates.

One might be confused by this last statement. Surely, one might say, biologists were classifying creatures long before anybody knew that taxa were monophyletic groups? Early biologists may have thought lice were generated out of dust, but they still classified creatures into phyla, classes and so on. Moreover, biologists were classifying creatures with a fair degree of accuracy long before the theory of evolution brought with it the idea of common descent. The discovery that these taxa were monophyletic groups was just that, a *discovery*. It was not extracted from the definition of the words phylum, class and so on. Does this not mean that the universal truth "If x is a salientiate, then x shares a common ancestor with all and only salientiates" is not true by definition after all?

No, it does not. Early biologists observed strong patterns of similarity among organisms, patterns of similarity that really do exist. On this basis they made their, often accurate, classifications. But modern biology tells us that, no matter how much a creature looks like a frog, if it does not come from the right ancestor then it does not belong in Salientia. Modern biology also tells us that it is massively unlikely that two separate monophyletic groups would be so similar as to be indistinguishable, but that is beside the point. In principle, if it turned out that some creature was indistinguishable from a frog in all its current characteristics, but was not descended from the common ancestor of salientiates, then it would not be a salientiate. How we would actually know this is hard to tell, but that too is beside the point. In practice, earlier classifications, and not just at the species level, have sometimes had to be redrawn because new evidence has come in (e.g. from gene sequencing) that showed what the true monophyletic groups were. Evidence that a creature does not have the right ancestry is evidence that it is not a salientiate, and trumps all other considerations that might tempt us to call it one. We could not discover that a salientiate was not descended from the common ancestor, because the discovery that a creature was not descended from the common ancestor would overthrow any presumption that it was a salientiate. It is at least possible in principle that we could discover that the speed of light in a vacuum sometimes varies. That is, the discovery that something that looked very like light travelled at a different speed in a vacuum would not force us to conclude that it was not light. In this sense, the generalization "If x is a salientiate, then x shares a common ancestor with all and only salientiates" is true by definition. This suggests that it is not a law.

10.1.4 Is there a law of natural selection?

What about the central idea of the theory of evolution itself, that is, that organisms evolve by natural selection? Earlier in this book I did not present this in the form of law; rather, I presented it as an explanation of why organisms are the way they are. Nonetheless, we expect *scientific* explanations to explain things by means of laws. The motions of the planets around the sun are scientifically explained by Newton's laws of motion. We might think, then, that there is a law of natural selection. How might we express such a law? I suggest something like this:

> *Differences in fitness in a population, plus selection pressure, will lead to an increase in the overall fitness in the population.*

But is this really a *law?*

That depends on how the term "fitness" is interpreted. It is tempting to interpret it as meaning something like strongest, healthiest and so on, but we need to make it more precise than this. As far as evolutionary explanation goes, fitness is whatever will increase an entity's[1] chances of surviving and reproducing in its environment. Whatever this is will vary depending on exactly what problems the entity faces. Being camouflaged obviously fits the definition, as it increases a creature's chances of evading predators, or of sneaking up on prey. But exactly what kind of camouflage depends on exactly what kind of creatures one needs to be camouflaged against. Different creatures have different sensory capacities: one creature has poor eyesight but good sense of smell, another vice versa. Moreover, creatures differ in what range of colour they can see. The kind of camouflage that is useful will also depend on what the landscape is like: a tiger's stripes make it harder to see in long grass, but would not be much use as camouflage in a neatly grazed field. This point can be made more general: the value of running fast, of having good eyesight, of being able to store water and so on will all depend on what problems a creature faces, which will in turn depend on the circumstances the creature finds itself in. There is no one thing that always enhances fitness. Let us see what happens when we plug the definition of fitness into the proposed law of natural selection. We get:

> *Differences in a population in whatever will increase an entity's chances of surviving and reproducing in its environment, plus selection pressure, will lead to an increase in the population in whatever will increase an entity's chances of surviving and reproducing in its environment.*

But this is starting to look like a roundabout way of saying:

> *If some entities in a population have what it takes to be more likely to survive and reproduce than others* [differences in fitness], *and not all the entities in that population can survive and reproduce* [selection pressure], *then the ones that have what it takes will be the ones that will survive and reproduce.*

And this, in turn, looks like something that is *trivially* true, or true by definition. The ones that survive and reproduce will be the ones that have what it takes, whatever that might be, to survive and reproduce. Looked at more closely, however, we can see that it is not actually trivially true. Suppose I had whatever it takes to win the World Chess Championship: superior skill in chess to anyone else, presumably. It would still not be true by definition that I *will* win the World Chess Championship. All kinds of things could happen; for example, I might fall ill, or suddenly lose interest in chess. Similarly, it is not true by definition that the entity that has whatever it takes to survive and reproduce *will* survive and reproduce. A creature could be the fastest runner, the strongest fighter and so on – that is, have more of *all* the traits that, for that creature in that environment, are useful for surviving and reproducing, than any of its fellows – but still be subject to any number of mishaps. So our proposed law is not a truth by definition.

However, the problem with it now is that it is not a strict law either, and this is for exactly the same reason as it not being a truth by definition. It is not only possible but *true* that mishaps sometimes befall the fittest individuals in any population. Moreover, the theory of evolution allows this, in the same way that it allows exceptions to Williston's law. It looks as though what we have, then, is a *ceteris paribus* law only, not a "proper" law. If we look back at my first formulation of the law we can see the same thing. It is not *always* true that differences in fitness in a population, plus selection pressure, will lead to an increase in the overall fitness in the population. Sometimes catastrophes happen, and populations go extinct instead of increasing in fitness.

Yet surely there is something importantly true in the proposed law. Moreover, surely it is central to biology in a way that Williston's law or the brain-mass law are not. Those other laws only apply to certain areas of life, for a start, whereas natural selection applies to life as a whole and as such. There is a sense in which it is exceptionless, in that there are presumably no creatures whose anatomy has not been shaped by natural selection through and through. Perhaps, then, we should make one more attempt to make it both strictly true and not just true by definition. We could modify the law to:

> *Differences in fitness in a population, plus selection pressure,* will tend to *lead to an increase in the overall fitness in the population.*

This is certainly at the heart of the theory of evolution, and is a principle that it, tacitly or explicitly, uses all the time to explain why creatures are adapted. But it does not get rid of the problem. The problem is now concentrated in the little phrase "tend to". On the one hand, it could be read in the same way as we might read the instances of "tends to" in Williston's law: the number of parts *tends to* decrease, while the diversity and specialization of different variations *tends to* increase. But that would just mean that it is a *ceteris paribus* law. On the other had, we could read it as meaning that in any such situation the ones who survive and reproduce will tend to be the ones who have what it takes to survive and reproduce. But then it would just be a trivial truth.

10.2 Does biology have real laws?

Throughout the previous section I have been assuming that neither *ceteris paribus* laws nor truths by definition can be genuine laws. Moreover, I have been assuming that the physical sciences – physics and chemistry – are in possession of genuine laws. However, it has been argued that such a picture of the physical sciences is unrealistic. Perhaps, contrary to what I have been assuming so far, the laws of physics and chemistry are either *ceteris paribus* or true by definition after all.

A case has been made by Mark Colyvan and Lev Ginzburg (2003) that at least a great many of what we accept as the "laws of physics" are in fact *ceteris paribus* laws. As an example, they give Galileo's law that "all massive bodies fall with constant acceleration irrespective of their mass". As they point out, there are plenty of objects that do not conform to this law: hailstones, snowflakes, parachutists and so on. So the law is *ceteris paribus* only. They anticipate the reply: but there are reasons why these cases do not conform to the law, which are well known. Snowflakes and the rest are sufficiently light that air resistance becomes a relevant factor in predicting their motion. The law, strictly speaking, only describes the effect of gravity on things, and therefore its predictions only come true if there is no other force involved. When there is no other force involved, it is true without exception. But Colyvan and Ginzburg reply to this that, in the case of many physical laws, the conditions they require are highly idealized, so that they are *never* in fact realized. Newton's laws of motion predict that the planets will orbit the sun in ellipses, but in fact they only do so in approximate ellipses, because of the gravitational effect of other planets, among other factors. So perhaps there is nothing wrong with *ceteris paribus* laws. Many of the laws that philosophers of science take to be the best exemplars of what laws should be like – that is, the laws of physics – are *ceteris paribus*. So the *ceteris paribus* laws of biology are just as much laws as any other.

Martin Carrier (1995) makes a similar point, with reference to Galileo's law. The fact that it has exceptions, he argues, should not be taken as meaning that it is not a law. Rather, it is a law that only applies when a specified set of initial conditions is met. The initial conditions include or imply that air resistance is not also acting on the object. Properly construed, the law should be read as "all massive bodies fall with constant acceleration irrespective of their mass, provided there is no wind resistance, and so on". So the fact that snowflakes and the like do not conform to the law does not stop it from being a law.

However, this argument may backfire. What Carrier's take on things suggests is that the law may actually be a strict, exceptionless after all, law and not a *ceteris paribus* law. Galileo's law holds true (or is supposed to) when the only force operating on an object is the gravity of the body it is falling towards. Even if we conceded that this is *never* the case – that even a falling bowling ball is influenced to some tiny degree by the air it is falling through, or by the moon's gravity, or any number of things – we could still say that, when the required conditions are met, the law applies *without exception*. We can say this even if the required conditions are *never* met, because we can at least *imagine* them being met. One might be tempted to think that the situation with Galileo's law is the same as that with the proposed law of natural selection. There is an open-ended list of things that might interfere with an object's descent towards the ground: air resistance, magnetic fields, passing birds and so on. Similarly, there is an open-ended list of things that might interfere with the level of fitness increasing in a population under selection. We can summarize the list with the labels "chance" and "constraint", but either of those terms can cover a multitude of possibilities. So, we might conclude, both laws are in the same boat, both being subject to an indefinite number of possible interfering factors. However, the two cases are still not the same. We can imagine – and in fact we can even construct a mathematically precise model of – a situation where nothing other than gravity acts on a body. We might be tempted to think that we can similarly imagine a situation where nothing but natural selection acts on a population. But can we?

The problem, I suggest, is that natural selection can *only* act where there is constraint. That is, not only is constraint always there, but it is impossible to conceive of natural selection taking place without it. This is, I suggest, for two reasons. First, to imagine a situation without constraint would be to imagine creatures with no constancy of structure whatsoever. We saw in Chapter 1 that heredity – that is, that offspring reliably resemble their parents to a high degree – is a necessary condition for natural selection. But this also constrains natural selection. Secondly, without constraint, there would be no situation to which creatures could become better adapted. I already argued this (albeit briefly) at the end of Chapter 4. Put simply, a problem arises for a creature because of *both* its environment *and* the way the creature is itself constituted. So the way the creature is constituted partly determines what would count as becoming better adapted, or fitter. But the way the creature is constituted also imposes constraints on how it can evolve. So, whereas we can imagine Galileo's law operating without any interfering factors, we cannot imagine natural selection operating without any interfering factors. Galileo's law is exceptionless in that it applies in every situation where gravity is the only thing operating. The proposed law of natural selection, by contrast, cannot be exceptionless in this way, because the very idea of "nothing other than natural selection operating" makes no sense.

Sober (1984: ch. 2) proposes a different approach to showing that there are real biological laws. Instead of trying to show that *ceteris paribus* laws can be real laws, he proposes that truths by definition can be real laws. As an example he offers: water is H_2O. He says much the same things about this as I said about salientiates above. The discovery that water is H_2O was an empirical discovery, and could not have been made just by looking at the definition of the word "water". But water is H_2O *by definition*, in just the same way that every salientiate shares a common ancestor with all and only salientiates by definition. That is, if we were to discover that some sample of a liquid was not H_2O, then we would have discovered that it was not water, no matter how much it looked and behaved like water. So, Sober, argues, something can still be a law even if it is true by definition.

This point can be related (although Sober does not do this) to a more wide-ranging point made by W. V. Quine in his famous paper "Two Dogmas of Empiricism" (1961). Throughout this discussion so far, I have been assuming that there is a difference between things that are true by definition, and things that are not. In philosophy this difference is commonly referred to as the *analytic–synthetic* distinction. As these terms are traditionally understood, *analytic* truths are those that are true just in virtue of definitions: "all bachelors are unmarried" is an example. *Synthetic* truths contain information that goes beyond what is contained in the definition: "John is a bachelor" is an example. One way in which one might clarify this distinction would be to say that in order to know that something is an analytic truth, all one needs to know is the definition. One can know that "all bachelors are unmarried" is true just by knowing what the words in the statement mean. But one could not know that John is a bachelor this way: one would have to know some facts about the world – that is, whether or not John is married – in addition to knowing what the words mean. Quine, however, argued against this distinction. One way we can understand what Quine is saying is to look again at the statement "water is H_2O". On the one hand, that water is H_2O is something that had to be discovered, and could not have been determined just by knowing the definition of the word "water". But on the other hand, if we found that something that looked just like water was not H_2O, we would say it is not water, rather than that we have discovered that some water is not H_2O. So, what exactly happened when scientists accepted that water is H_2O: was an empirical discovery made about water, or was the definition of the word "water" changed? We would surely have to answer: both. The definition of "water" changed, but *because* of empirical discoveries. So, Quine is urging, the neat distinction that we make between truths by definition and other kinds of truths cannot be made. But there is more still. There could conceivably be some revolution in science in the future such that scientists no longer thought that the stuff we drink, swim in and so on was H_2O. This would mean that we would no longer say that anything that was water has to be H_2O. Something that we now treat as true by definition would have turned out to be false. If we accept Quine's claim, then, it does not matter whether biological laws turn out to be true by definition or not, for there is no real difference between a truth by definition and some other kind of truth.

Quine's view is, admittedly, still controversial. Nonetheless, the upshot of our discussion of whether biology has laws may just turn out to be: it depends on what you mean by a law. If you agree with Quine, then something's being true by definition is no obstacle to its being a law, since there is no clear distinction anyway. One does not have to go as far as Quine to agree with Sober, provided one thinks that something that is true by definition can be a law. If one does not agree with Sober on this, then one may have difficulty, not just accepting biological laws, but accepting physical laws as well. I started this chapter by remarking that philosophers of science tend to take the physical sciences as exemplary. If they are right, then biology needs to be either shown to be like physics in the relevant respects, or else accepted as being an inferior science. I want to argue against this in the next section: in other words, I want to argue that biology is different in important respects from the physical sciences, but in no way inferior.

10.3 Comprehensiveness, unity and simplicity

The theory of evolution has been claimed to be a great, perhaps even the greatest, success story in science: "biology, unlike human history or even physics, already has its grand unifying theory, accepted by all informed practitioners, although in varying versions and interpretations" (Dawkins 2004: 8). This claim sounds, and I think Dawkins intended it to sound, both surprising and provocative. Physics is extremely impressive in its success in uncovering laws that apply to the entire universe and can apparently be reduced to a very small number of mathematical formulae: "just six numbers", according to Martin Rees (1999). This is not the place to assess the claim that physics lacks a grand unifying theory. What I want to address is the claim that biology has one.

What do we mean, anyway, when we say that a theory is unified? One thing we might mean is that it is *comprehensive*. There are well-known episodes in the history of science where an old theory is replaced by one that is more comprehensive. Newton's laws of motion worked well for many cases: those where the speeds involved did not approach that of light. It was replaced by Einstein's theory of relativity, which gave all the correct results that Newton's theory did, but also gave correct results in cases where Newton's theory got it wrong.[2] The theory of relativity is a more comprehensive theory.

Presumably there is an ideal of a theory that is *maximally* comprehensive: that is, it successfully covers all the phenomena that it is supposed to cover. Newton's own laws of physics did quite well in this regard, bearing in mind that some of the phenomena predicted by Einstein's theory but not by Newton's were not known about in Newton's time. Nobody knew that stars would appear displaced during solar eclipses, for example. Nonetheless, even before Einstein came along, there were facts that scientists had difficulty accommodating into Newton's theory. The apparent orbit of the planet Mercury, for example, did not match calculations based on Newton's formulae. In that sense, then, Newton's theory failed to be maximally comprehensive. Can we say, then, that biology is maximally comprehensive in the sense of covering all the phenomena it is supposed to cover?

10.3.1 Comprehensiveness and spatiotemporal boundedness

One way in which one might argue that biology is not comprehensive in the way that the physical sciences are is that biology only covers a highly restricted domain. The laws of physics cover the entire physical universe. But biology only covers living things, which only make up a very small part of the universe. So, one might argue, biology not only is not, but could not possibly be, comprehensive in the way that physics is. One might reply to this that there is no reason to expect biology to be comprehensive in that sense. I defined a comprehensive theory above as one that successfully covers all the phenomena that it is supposed to cover. Biology, then, cannot be said to fail in this regard if it fails to cover the non-living universe, can it? We do, after all, have physical sciences that deal with only part of the physical world. The science of fluid dynamics deals with the motions of liquids. We do not say it fails to be comprehensive just because the universe is not entirely composed of liquids. It covers all it is supposed to cover. Why can we not say the same about biology?

This is not the final answer, however. The sense in which fluid dynamics deals with a restricted domain is not the same as the sense in which biology deals with a restricted domain. If the models of fluid dynamics are accurate, they are accurate as models of how liquids move *anywhere in the universe*. The category of "liquid" is spatiotemporally unbounded in the sense that I outlined in Chapter 9. Fluid dynamics, it might be said, does not just deal with all the liquids we happen to know about, or even all the liquids there happen to be, but with *liquids as such*. By contrast, biology deals with a phenomenon that is spatiotemporally bounded. Life, as far as we know, only exists on earth. Moreover, as I also pointed out in Chapter 9, the different species are spatiotemporally bounded. Each species began at a particular (albeit vaguely defined) time and, if it comes to an end, will never exist again (at least, barring *Jurassic Park* scenarios). But the same is true, it might plausibly be claimed, of larger biological kinds as well. Chordates came into existence at a particular time, and if they all went extinct there would never be any chordates again. Moreover, if creatures that looked very like chordates evolved separately on another planet, they would not be chordates, not having the right ancestry. We can push this point even further and say that life itself began at a particular time, and hence is spatiotemporally bounded too. So, while fluid dynamics deals with something that could crop up anywhere in the universe, biology does not.

But the last step in this argument is, as philosophers sometimes say, too quick. Chordates, and all other biological categories, are defined according to their ancestry. If you do not have the right ancestry, you do not belong in the group, no matter how similar you are to the creatures in the group. But that does not mean that that is true of the category "living thing" itself. We all accept that it is at least conceivable that there could be life on other planets. It would be very strange to discover things on another planet that looked and behaved like living things but deny them the name "living thing" because they are not descended from thermobacteria, or whatever we are all descended from. Doubtless, our biologists would get busy inventing new taxonomic names for these extra-terrestrial creatures, assuming extraterrestrial biologists had not already done so. And they might want to invent a taxonomic name for terrestrial living things as a whole. We would deny *that* name to the extraterrestrials, but we would not deny the name "living thing" to them. Of course, we might have difficulty recognizing that extraterrestrial entities are alive, but we would need a better reason to deny that they are, than that they are not related to us.

Recall Dawkins's argument to the effect that if there is life anywhere in the universe, it has come about by natural selection (summarized in Chapter 4). He also said that "functional complexity" is "diagnostic for life". Whatever the merits of these particular claims – and Dawkins makes a good case for them – we can at least say that there is a difference between living and non-living, even if the boundary is vague and we cannot define exactly what life is. Moreover, that difference is in principle applicable to entities anywhere in the universe. The fact that we understand what it means to say that there could be life on other planets shows that we all accept this. The only account we have for how life could come about is an evolutionary one, one in which – as even Gould and Lewontin accept – natural selection plays an important, central part. We can say with a fair degree of confidence, then, that the theory of evolution is spatiotemporally unbounded in its application, in exactly the same way as fluid dynamics is. It applies to *life as such*.

10.3.2 Unity and pluralism

In Chapter 5, I argued for pluralism on the units of selection question. As we saw in Chapter 2, it has been argued that the gene-selectionist view of adaptation is more comprehensive than Darwin's own view, which focused on the survival and reproduction of organisms. But, as we also saw, the stronger claim is sometimes made that the gene-selectionist view is "always available" as a way of spelling out any adaptationist story. However, as we saw in Chapter 5, developmental systems theory makes a counter-claim; namely, that it covers cases that the gene-selectionist view fails to cover, and is itself capable of covering all cases of adaptation. Thus, we have two rival theories, each of which is claimed to have maximum comprehensiveness. However, we saw in Chapter 3 that a multiple-level view of selection seems plausible to many philosophers of biology today.

Similarly, in Chapter 9 we saw that there is a good case for pluralism in how we decide whether a group is a species. The complications of ring species, hybridism and the fact that the interbreeding criterion cannot be applied to asexually reproducing creatures all suggest that there is no single criterion for deciding if a group is a species. Either there are cases – and a great many of them at that – where we cannot call a group a species at all, or we have to use different criteria in different parts of the tree of life. So it looks as though any account of what makes a group a species will also fail to be maximally comprehensive.

Contrast these situations with analogous situations in the physical sciences. It could turn out that group selection is needed to account for some adaptations. But it could not possibly (and nor is it claimed to) account for all of them. Gene selection – or possibly individual organism selection – will still be needed. But now imagine a situation in physics where there are two theories, A and B. They agree in their predictions on a very large range of phenomena, just as Newton's theory and the theory of relativity do (or very nearly do). Moreover, in most of those cases where they agree they fit the observed phenomena. But, in the real-life case of Newtonian mechanics and the theory relativity, when they disagree Newton gets it wrong and the theory of relativity gets it right. Imagine instead, that where theory A and theory B disagree, theory A sometimes gets it right where theory B gets it wrong, *and sometimes vice versa*. If this were the case, we would have to say that *neither* theory A nor theory B was maximally comprehensive. In each case, there would be known facts that did not fit the theory. This is an unsatisfactory situation. Something like this was the situation in nineteenth-century physics, where there were two rival theories of light, the wave theory and the particle theory. Each theory posited entities whose existence was not recognized in the other. Physicists, it seemed, needed to tell stories wherein light is a wave, *and* stories wherein light is a particle, in order to cover all the facts concerning light. There was no viable theory that held that some light is waves and some light is particles.

Now try to imagine a situation in the physical sciences analogous to the situation with regard to the definition of species. We have clear criteria for deciding whether something is gold or any other of the chemical elements. If it has the atomic number 79 it is gold. The difference that I am pointing to here between this and the criteria for specieshood is not that the latter are vague, although that is also a significant difference. Rather, the difference is that the same criterion is used *in every case* where we want to decide what element something is: the atomic number only. This is not the same as saying that that is the *empirical method* for finding out whether something is gold or lead or whatever. Rather, it is what *makes* something gold or lead or whatever. To put it another way, two atoms are the same element if and only if they have the same atomic number. Something is an element if and only if it has *an* atomic number, so water, milk and fire are not elements. Imagine if, instead, we had a criterion for deciding whether something was a sample of element X, and a different criterion for deciding whether something was a sample of element Y. Something, say, is a sample of element X if and only if it boils at 300 °C at 1 atmosphere pressure; but something is a sample of element Y if and only if it has a specific gravity of

3.5. Of course, either of these things could conceivably be the empirical method we use to decide if something is this or that element. But such a pluralism of empirical methods would not mean a pluralism of criteria for what makes something the element it is. If we did actually have such a pluralism that would be, once again, an unsatisfactory situation. But, when it comes to species, the pluralism of criteria for deciding what species some organism is, and for deciding whether a group is a species, is a pluralism *that goes all the way down*. That is, there are not just different empirical methods in different cases for deciding what species something is, there are different criteria in different cases for what makes something a member of a particular species. Note, too, that this is not a situation that we should expect some future perfected biology to rectify. Does this mean, then, that biology is not a unified science after all? And is it a cause for complaint?

The idea that older theories are revealed to be mere approximations by newer theories that encompass everything the old ones do, and more, owes its appeal, I would suggest, to the way it seems to fit at least some of the major events in the history of physics.[3] Whatever its applicability to physics, it looks as though attempts to apply this picture to biology (or at least to this question in biology) are mistaken. Consider a different analogy than the one from physics. Many historians reject Marx's claim that all of culture can be seen as a mere superstructure on the fundamental realities of means of production. This is not just because those historians are reactionaries, but because they perceive that many salient historical facts simply do not fit this picture. More generally, historians nowadays are often reluctant to attempt to explain all of history – even its broad patterns – in terms of any single factor. Further, such monistic pictures of history tend to perform very badly at making sense of developments that take place after the theory is devised. I propose, then, that the sum total of adaptive explanations in biology is unified *at least* in the way that a historical narrative is unified: the existence of different driving forces, different explanatory factors, does not mean that we have incompatible stories coexisting. The answer to the units of selection question, then, may turn out to be like the answer to the question: what is the driving force of history? That is: many things, and in the future it may be different things.

Even the possibility of there being more than one unit of selection points to a disanalogy with allegedly comprehensive theories in physics. For we expect a comprehensive theory in physics to continue being comprehensive in the future as well as being comprehensive in the past and the present. As an alternative scenario to a new type of replicator emerging, imagine that we discover some type of replicator of which we had no previous knowledge. This would not bring about a revolution in biology, but merely be one more item to add to the list of units of selection that we already had (this would be so even if that list had previously consisted of only one item). By contrast, imagine that some new, more comprehensive, physical theory of motion emerged that covered all the phenomena that the best current theory (which I have been presuming to be the theory of relativity) covers, plus more. That *would* be a revolution in the physical sciences. Even if we had some finite list of all the types of replicators there are, the reason why *that* was the complete list would not be any deep fact about biology, but merely the ins and outs of evolutionary history as it has so far played out. So there is no deep answer to the question: given that something is an adaptation in so far as its presence is explained by the contribution it makes to the propagation of x, what is x? There is only the trivial answer: whatever replicators there happen to have been, up to this moment.

Another way to put this point is to say that the different perspectives taking different items as the replicator are *not* different paradigms, as the wave theory and the particle theory of light were. For the wave and particle theories each respectively posited entities whose existence was not recognized by the other. There was no viable theory that held that some light is waves and some light is particles. But a theory that focuses on one replicator is not committed to denying the *existence* of the entities that different theories take to be the replicator. A gene-selectionist view, for example, does not deny the existence of organisms or groups. Rather, it makes the empirical claim that adaptations cannot come about that serve the interests of something other than genes. To explain some features of organisms by x-selection and others by y-selection is not evidence of radical disunity in our biological theories, in the way that explaining some features of light by its being a wave and others by its being a particle, was evidence of a radical disunity in our physics. In that sense, the futility of the search for a comprehensive answer to the question of the replicator does not mean that biology lacks unity in any way that need cause us concern.

To understand this point, compare the following two scenarios:

(i) We need to tell stories wherein adaptations are explained by their contribution to the propagation of xs, *and* stories wherein adaptations are explained by their contribution to the propagation of ys, in order to cover all the facts concerning adaptation.

(ii) We need to tell stories wherein light is a wave, *and* stories wherein light is a particle, in order to cover all the facts concerning light (the situation in nineteenth-century physics).

Scenario (ii) constituted a crisis in the science, because the two types of story were *ontologically incompatible*. To resolve the crisis required a revolution in the science. Scenario (i), in contrast, does not constitute a crisis in the science, because the two types of story are ontologically compatible. A theory that focuses on one replicator is not committed to denying the *existence* of the entities that different theories take to be the replicator. A gene-selectionist view, for example, does not deny the existence of organisms. Rather, it makes the claim that adaptations do not come about that serve the interests of something other than genes at the expense of genes. While there could not be a viable theory that held that some light is waves and some light is particles, an account of adaptations that takes some to be for the propagation of xs and some for the propagation of ys is perfectly viable. An account of selection that has more than one type of replicator can still be unified in the sense of being continuous. That is, every part of the grand evolutionary narrative can (at least in principle) be connected to every other part.

It might be felt that ontological compatibility is not sufficient to ensure a non-crisis situation in the science, or to motivate my claim that we may as well call off the search for a comprehensive, unipolar account of natural selection, or a single definition of "species". It might be asked: what about the consideration of simplicity? Might this not motivate a search for a unipolar account of selection or a single definition of species? I will briefly say something about simplicity.

Simplicity is often spelt out in terms of Ockham's razor. The principle is that entities must not be multiplied without necessity, and "without necessity" is crucial here. We can take it to mean "without going beyond what's needed to explain *all* the phenomena". There may well be a necessity for more than one type of replicator to explain all adaptations. If that's so, Ockham's razor would not be violated.

One can again make the comparison between the evolutionary narrative of biology and history. Many historians think single-factor accounts are insufficient to cover all the facts. We would not think a historical account was violating some methodological principle of simplicity just because it appealed to more than one explanatory factor.

10.4 Conclusion

The fact that biology lacks strict laws of its own, and the fact that many of its core concepts are vague and have plural definitions, show that it is a different kind of discipline from the physical sciences. In a sense, we might say that science itself is not unified. It is important to be clear about just what that sense is, however.

In the mid-twentieth century, one of the questions that preoccupied philosophers of science was known as the *demarcation* question. It was widely perceived that a number of disciplines that were not sciences were trying to pass themselves off as such. Popper was one of the people who felt this most strongly, and devoted three books to the task of formulating a criterion for distinguishing genuine science from pseudo-science. I will not discuss the merits of Popper's demarcation criterion here, but merely note that his star example of good scientific practice was Einstein's theory of general relativity, and the test carried out by Eddington to corroborate it that I mentioned in Chapter 1. He was convinced before he developed his criterion that the theories of Freud, and those of Marx, at least after Marx's followers had revised them, were pseudo-sciences. So what he desired was a criterion that would admit Einstein but would keep Freud and the Marxists out. Of course, he also wanted to keep the really obvious examples of pseudo-sciences, such as astrology, out, and keep the really obvious examples of genuine sciences in. But there's the rub. Most people would agree that biology is a science, and that the theory of evolution is a perfectly respectable part of biology. But Popper had difficulty making the theory of evolution fit his criterion. He concluded that it was not a scientific theory but a "metaphysical research programme" (Popper 1976: 167–80). (Incidentally, he did not say that it was not *true*.) His basis for saying this seems to have been that it was not testable, not having, as he correctly observed, strict laws. Whatever the merits of this claim, it seems to stem from a persistent tendency of philosophers of science to take the physical sciences as the model of what all sciences should be. I hope I have convinced you by now that biology is just not like that, and should not be expected to be.

A number of current philosophers of science argue for the view that science is radically lacking in unity. Thus, we have books with titles such as *The Disorder of Things* (Dupré 1993) and *The Dappled World* (Cartwright 1999). Both of these philosophers call for an abandonment of the idea that there is one right methodology for all of science, and consequently cast doubt on the idea of a criterion for demarcating science from everything else. For example:

> Science, construed simply as the set of knowledge-claiming practices that are accorded that title, is a mixed bag. The role of theory, evidence, and institutional norms will vary greatly from one area of science to the next. My suggestion that science should be seen as a family-resemblance concept seems to imply not only that no strong version of scientific unity of the kind advocated by classical reductionists can be sustained, but that there can be no possible answer to the demarcation problem.
>
> (Dupré 1993: 242)

This is not quite as pessimistic a conclusion as it sounds. It does not imply, for example, that we have to say that astrology is as good as physics. It just means that whatever reasons we have for saying that it is not, do not arise from some criterion that also performs the task of separating science from non-science. If we want to say (as I certainly do) that physics and biology are good, while astrology is bad, that purpose is not served by saying that physics and biology are sciences but astrology is not. Ultimately, Dupré concludes that "science is a human product, … [and] like other human products, the only way it can ultimately be evaluated is in terms of whether it contributes to the thriving of the sentient beings in this universe" (*ibid.*: 264).

We need not go as far as Dupré does here, saying that that is the *only* way. Surely, it might be thought, we can evaluate theories by looking for reasons to think they are true or false, such as: do they fit the evidence? Are they consistent? We can still say that that does not demarcate science from non-science because we use evidence, considerations of consistency and so on in disciplines that are not usually called science, such as history, biography and philosophy, and for that matter in our everyday lives. I take it that the core theories of physics and biology, as well as much of what is in the history books, and other things, pass those tests, and that astrology does not.

Cartwright looks at arcane areas of physics, and argues, apparently on the basis of the same disunity there that Dawkins alludes to, that we should just accept that it is disunited. To me this seems a rather pessimistic conclusion, ruling out the possibility of a unified physics in the future. Be that as it may, the disunity of physics does not carry over into biology. Biology takes the laws of physics as a given, and explains how we get living things out of them. Hence, whether physics eventually achieves its grand unified theory or not does not affect biology. The theory of evolution is an important step on the road to an *ontologically* unified scientific world-picture. That is, the theory of evolution gives us an account of how living things come into existence that does not require us to posit any laws or entities over and above those posited by the physical sciences.

If one were truly adamant that science should aspire to strict, exceptionless laws, and concepts with single definitions, one could use the arguments in the previous sections to conclude that we should simply replace biology with the physical sciences, and get strict laws that way. But even if that were possible, we would not have *explained* the phenomena of biology. We would have rendered them invisible.

11. Evolution and epistemology

The theory of evolution is often taken to be a triumph for *naturalistic* conceptions of the world. That is, it is often taken as a powerful vindication of the view that there is nothing that, in principle, cannot be explained naturalistically using the resources of the empirical sciences. Part of the general programme of many naturalistic philosophers is to replace traditional philosophical questions with scientific ones. This approach to philosophy was explicitly endorsed by John Dewey and W. V. Quine, both of whom have proved to be extremely influential in this regard. The questions of epistemology, such as "What is knowledge?" and "What can we know?" are among these traditional philosophical questions. There are two ways in which we might try to accommodate philosophy to a naturalistic worldview:

- It can be held that instead of asking purely conceptual questions, such as "What is knowledge?", we should ask only empirical questions, such as "How, in fact, do human beings come to have the beliefs they have?" This approach, however, may be considered to be sidestepping the questions of traditional epistemology, rather than using science to help answer them.
- There is a more modest way of going about things. Whatever answer we give to questions such as "What can we know?" or "How is it possible that we can have knowledge?" had better be compatible with what our best current science tells us about what capacities human beings have. On this view, the question "What is knowledge?" can remain as a conceptual question, but the question "What can we know?" is, at least in part, empirical. If we want to find out what human beings can know, then scientific findings about our neurophysiology, our cognitive capacities and so on, may be relevant. Evolution may be relevant, since it may tell us about what capacities we are likely to have. Certain capacities, it might be held, are *the kind of thing that evolution is likely to produce*, while others are not. To

take an extreme case, evolution could not produce a species that had no means of securing its continued existence in future generations. Perhaps, then, evolution can tell us about capacities that human beings are likely to have that will in turn shed light on traditional epistemological questions such as "What can we know?" I shall be concerned here with this second, more modest, naturalistic type of project.

11.1 Conjectures and refutations

Karl Popper is probably the most famous name in philosophy of science, and is often invoked by scientists when they want to label some theory "pseudo-scientific". I shall very briefly outline his philosophy of science here, before moving on to the specifically evolutionary aspect of his views.

Popper proposed a solution to the *problem of induction*. First formulated by David Hume ([1739–40] 1978: bk I, pt III, §II), the problem of induction is this: suppose you have seen a thousand swans, all of which have been white. Is it rational to infer from this that all swans are white? Hume says no, because what you have seen only tells you that all the swans you have seen were white. It is invalid to infer from this that "all swans are white". Just in case it is not obvious why, think of a *deductive* inference: all men are mortal, Socrates is a man, therefore Socrates is mortal. What makes this a valid inference is that if the premises are true the conclusion *must* also be true. However, as Hume pointed out, this is not the case with an inductive inference. Even if every swan you have seen is white, it could still be the case that not all swans are white. Hume concluded that induction is fundamentally non-rational; it is just a habit or instinct that we have.

Popper's solution was to point out that, although we cannot prove a universal claim *true* by means of examples, we can prove one *false* by means of examples. For example: swan X is black, therefore not all swans are white. Thus, rejecting a universal claim could be rationally justified. Popper wanted to show that, *contra* Hume, science, and indeed everyday thinking in general, could be rational. He rightly perceived that in both science and everyday life we make universal generalizations: water boils at 100 °C at 1 atmosphere pressure; light always travels at 299,792,458 metres per second in a vacuum; apples are edible; wasps sting. If we tried to get along without such generalizations, we would have extremely impoverished science, and lives. Since we can only be rationally justified in rejecting a universal generalization, what we should do is make generalizations and hold on to them until they have been falsified. This argument gave rise to Popper's famous *falsificationist* criterion for separating real science from pseudo-science: real science makes falsifiable claims, pseudo-science does not.

The demarcation issue need not detain us here: what I am concerned with is Popper's claim about how knowledge actually develops, and how he thinks that can be shown to be, at least when it is done properly, a rational process.[1]

According to Popper, in both science and everyday life we habitually make "bold conjectures", that is, we make hypotheses that are falsifiable. A child may make the bold conjecture "if I put my hand in the fire, nothing bad will happen". She may test it, and find very quickly that it is false. Or she may make the bold conjecture "if I make the sound 'bikkie', mummy will give me one of those round, tasty things". She may test it, and get a biscuit, but this does not prove the conjecture is true; it only fails to prove it is false. On Popper's terminology, the second hypothesis has been *corroborated*. The child says to herself "so far, so good" and provisionally accepts the hypothesis as true. Later, she may make the sound "bikkie" and not get a biscuit. When this happens, she might try a different bold conjecture "if mummy is frowning when I make the sound 'bikkie' she will not give me one, but if she is not she will". This can then be tested and corroborated or falsified. In this way, we continuously replace hypotheses about the world with new ones. The new ones are better because we have eliminated errors. Thus, according to Popper, knowledge progresses.

Popper explicitly calls this theory of knowledge an evolutionary one, and elaborates on the parallel with biological evolution in Chapter 7 of *Objective Knowledge* (1979). The analogy between the above account and natural selection seems clear enough. Organisms are well adapted because they have survived the vicissitudes of life that could kill them or prevent them from reproducing. Natural selection is a continuous process of eliminating designs in favour of better ones. Our beliefs, according to Popper, are good because they have survived the vicissitudes of experience that could prove them false. The development of knowledge is a continuous process of eliminating hypotheses in favour of better ones. Moreover, one of the consequences of Popper's view is that it does not matter how a theory originates; all that matters is that it can stand up to the test of experience. This can be seen as analogous to the Darwinian claim that mutations are random; all that matters is that the new features produced can stand up to the test of living. In principle, a complete scientific novice could come up with a new, good, theory. All that makes this unlikely is that the demands that a theory has to meet are very stringent. Similarly, in principle a mutation could produce a new, adaptive, trait. All that makes that unlikely (remember that by Darwin's logic most mutations are either harmful or make no difference) is that the demands that an organism has to meet are very strict. In both cases, what counts is not how a new theory/trait originates; what counts is how it stands up to tests. The process of generating new theories/traits does not have to be in any way directed. It is the continuous weeding-out of bad traits that guarantees the adaptedness of an organism. It is the continuous weeding-out of bad theories that guarantees the progress of science.

Tempting though the analogy is, however, it falls down on some points. The last sentence in the previous paragraph contains a clue to one of the problems. *Progress* in science presumably means getting closer to the truth. Even though Popper thinks we will never be able to say that we have reached the final theory (even if we had a science that was completely true and comprehensive, how could we know this?), he thinks that newer ones are closer to the truth than older ones, in that they have eliminated the errors of the older ones. But, presumably, this is supposed to mean *objectively* closer to the truth. For example, Popper would not claim that Ptolemaic astronomy was true in the middle ages but not in our time; he would say it was false at all times. The knowledge that we seek is *timeless* knowledge. If we are to talk of progress, we at least want to say that science is getting better by some *universal* standard. By contrast, a creature that is well adapted to one environment may be badly adapted to another environment. Two creatures may be completely different, but both well adapted. We should not think of evolution as progressive in the strong sense of leading towards outcomes that are objectively, universally better. Some evolutionists claim that it is progressive in milder senses: that there are long-term trends towards greater complexity, for example (think of Williston's law). But even if that is so, we can still only speak of creatures being adapted to specific circumstances. There is no such thing as being "objectively", timelessly, well adapted. If our view of the development of knowledge was to be analogous to evolution, we would have to be *relativists*, that is, we would have to say what is true at one time and place is false at another, and vice versa. Popper abhorred this kind of relativism.

There is also a problem with the analogy between the falsification of hypotheses and the elimination by natural selection of organisms, traits, species or whatever. Bear in mind that Popper wants to show that preferring one hypothesis over another can be rationally justified. Science progresses not just because theories are falsified, but because they are *rightly judged* to have been falsified, and consequently *rationally* rejected. By contrast, the elimination of something by natural selection can be explained by completely blind, mechanical forces. To be analogous to natural selection, the rejection of theories would have to be something that happens even without there being rational grounds for it. It is true, and Popper accepts, that the history of science does not always conform to his idealized picture. A theory may be rejected rationally because it has been falsified, or it may be rejected irrationally for any number of reasons. Perhaps the person who proposed it is not part of the scientific establishment, or is not good in public debates. Or perhaps the theory is offensive to many people. Both the rational rejection and the irrational rejection may be considered to be cases of the theory failing because it is not adapted to its particular environment. That environment *may* include an ideally rational scientific community, but it may also include prejudices against certain people, or unwillingness to accept views that are deemed offensive. But Popper wants to distinguish between these two types of cases. The analogy with natural selection does not give us any basis for doing so.

As an attempt to answer the questions "What can we know?" and "How is it possible that we can have knowledge?", Popper's account is disappointing. If the analogy with natural selection was the whole story about how knowledge develops, then we would have to say that there is no such thing as an objectively true theory, and that theories just come and go for many different reasons. Some people (e.g. Richard Rorty) do hold this view, but it was anathema to Popper. Popper wants to give an account that shows how we can be objectively *right* in our choice between two theories. But, if the analogy with natural selection is to be taken seriously, his account fails to do so.

Moreover, there is reason to believe that his account of how we build up knowledge is *empirically* false. According to Chomskian linguistics, cognitive science and evolutionary psychology, even very young children already have many presuppositions about the way the world is. Chomsky claims that we have inbuilt knowledge of the rules of universal grammar; and it seems well established that we have presuppositions about how physical objects behave, and many other things. Some of these inbuilt presuppositions may be false, but either way they are often highly recalcitrant to change in the light of experience. This does not seem to fit in with Popper's picture of a person generating and testing bold conjectures. I shall return to the subject of evolutionary psychology in Chapter 13.

11.2 The reliability of our sources

There is a further traditional epistemological question: what, if anything, guarantees the truth of our beliefs? This question dominated philosophy in the seventeenth and eighteenth centuries. The different approaches at that time – rationalism and empiricism – were both forms of *foundationalism*. That is, on both of these views there were some beliefs that were foundational in the sense of being guaranteed to be true, without needing any justification by other beliefs. Descartes, for example, held that he knew without justification that he was thinking. Empiricists usually held that one knows one's own sense-data in this direct way. Foundationalist epistemologists, then, attempt to show that we can justify other beliefs by deriving them from the foundational ones. This often led to odd claims about what we know and how we know it. Descartes held that he could know that there is a God simply from the fact that he possessed the concept of God, and that knowing that there is a God justified belief in an external world. Empiricists, on the other hand, had difficulty explaining how we can know anything but sense-data themselves. Historically, foundationalist epistemologies have often courted the danger of collapsing into scepticism. They give an answer to the question "What guarantees that our beliefs are true?" that only seems to work for a very small subset of our beliefs.

In the twentieth century, *coherentism* arose as an alternative to foundationalism. On this view there are no foundational beliefs. Rather, beliefs form an interconnected web. There are variants of the coherentist view, but on an influential version developed by Donald Davidson (e.g. Davidson [1974] 1984), any person's beliefs *necessarily* form an interconnected web, and must be mutually supportive, in that every belief is supported by other beliefs in the web. Moreover, according to Davidson, it is conceptually impossible to have a web of beliefs that is not largely true. Many people think Davidson's view is overly optimistic. Surely, it might be argued, a plausible fairytale would be coherent, but that would not mean it was true. Moreover, if beliefs are by definition mutually supporting, it is not clear how one would pick out the ones that are false. Historically, coherentist epistemologists have often courted the danger of collapsing into relativism. By making some minor adjustments to Davisdon's position, Rorty arrived at a view that is highly relativistic. Coherentism gives us an answer to the question "What guarantees that our beliefs are true?" that just seems implausible.

As a way out of this impasse, we might think of the question "What guarantees that our beliefs are true?" as similar to the question "What guarantees that creatures are well adapted to their environment?" The similarity can be made more apparent if we think of the notion of truth as somehow involving *fitting* or *being adequate to* the way things are. We might say: what makes a belief true is that it stands in some relation to some feature of the world. The belief "water is wet" is true because water is wet. That is, there is a feature of the world – the fact that water is wet – that the belief is appropriately related to, and it is this being appropriately related to it that makes the belief true. I am being deliberately noncommittal with use of the phrase "appropriately related"; I shall eventually say something about what the relation is supposed to be. For the moment, what I am trying to convey is the hopefully not too controversial thought that it is *something about the way the world is* that determines what beliefs are true. Different ways of expressing this might include "true beliefs describe the world accurately", "true beliefs correspond to the world" and so on.

What can we say about false beliefs? If true beliefs stand in an appropriate relation to bits of the world – whatever that relation is – then we can say that false beliefs fail to stand in that appropriate relation to bits of the world. They do not describe bits of the world accurately (or at least, not the bits they are supposed to describe), they do not correspond to bits of the world and so on.

By analogy, we might say that adaptive traits *fit* or *are adequate to* the way things are. We do speak, after all of a creature being adapted *to* an environment, and an environment is a bit of the world. So, a trait is adaptive if it stands in some appropriate relation to a bit of the world. Conversely, a trait is maladaptive if it fails to stand in that relation.

Once again, this talk of "standing in appropriate relation to" is rather vague. But we can make it more explicit. A trait is adaptive if it helps the creature that has it to survive and reproduce (or if it helps its genes replicate or what have you but, as I argued Chapter 3, most of the same things will be adaptive whichever account you go for). We have to add: given the way its environment is. Traits, we might say, latch on to features of the environment. It is often said that when an environment is changing, natural selection "tracks" changes. If the environment is very dry, then a trait that helps the creature store water, such as a camel's hump, is likely to be adaptive. If the environment contains a dangerous predator, then the ability to detect and evade it is likely to be adaptive. We can give a simple explanation for why organisms are generally well adapted to their environment: if they were not, they would have gone extinct. Natural selection has made them that way.

Our epistemological question was: what guarantees the truth of our beliefs? This brief discussion suggests an answer: by and large, it is useful to have true beliefs, and harmful to have false ones. Here are a few examples. Suppose one cave person has the belief "lions are not safe to approach", while another has the belief "lions are safe to approach". We can be fairly sure which belief would be weeded out by natural selection. Another widely used example is: suppose a cave person sees three bears going into a cave, and then sees two bears coming out. The person reasons: the bears have left the cave, so it is safe to go in. The knowledge that $3 - 2 = 1$ would clearly have been useful. Hence, the argument goes, basic mathematical reasoning ability is likely to have been favoured by natural selection.

Hofstadter (1985: 577–8) gives a more subtle example. We make generalizations all the time, and we need to do so, as we could not possibly treat every situation as if it were completely new. But to be able to make generalizations we need to be able to recognize one thing as falling into the same category as another thing. Only if we recognize lions as forming a class are we able to have the useful belief "lions are not safe to approach". But there are many different ways in which we could classify things. Suppose two cave people are already familiar with lions and with zebras, and know which are safe to approach and which are not. One day they see a tiger. One cave person decides that the tiger fits into the same class as a lion, because it is a similar shape; the other decides it fits into the same class as a zebra, because it has stripes. So one infers that it is not safe to approach and the other that it is, with predictable results. Thus, Hofstadter suggests that the ability to pick out the relevant similarities between things – as Plato says, to "carve reality at the joints" – is a product of natural selection.

So the answer to our question "What guarantees that our beliefs are true?" is: natural selection would weed out those who have false beliefs. This line of reasoning is popular with evolutionary psychologists, but perhaps the classic statement of it is by Quine, in an essay called "Natural Kinds":

> [W]hy does our innate subjective spacing of qualities accord so well with the functionally relevant groupings in nature as to make our inductions tend to come out right? Why should our subjective spacing of qualities have a special purchase on nature and a lien on the future?
>
> There is some encouragement in Darwin. If people's innate spacing of traits is a gene-linked trait, then the spacing that has made for the most successful inductions will have tended to predominate though natural selection. Creatures inveterately wrong in their inductions have a pathetic but praiseworthy tendency to die before reproducing their kind. (1969: 126)

Two conclusions can be drawn from this evolutionary reasoning.

(i) In so far as we have any innate beliefs (putting a suitable gloss on "innate"; see Chapter 6), they are highly likely to be true, as they are highly likely to be products of natural selection.

As it happens, evolutionary psychologists have a number of arguments to the effect that we have innate (or at least evolved, developmentally robust) cognitive apparatus, which includes things that we could reasonably call beliefs. We have, for example, the belief that solid objects cannot pass through other solid objects, that water quenches thirst, that animals move under their own power without being pushed, and so on. I shall leave these arguments for Chapter 13. Let us suppose that you do not find the idea of innate beliefs – on any reasonable glossing of the word "innate" – plausible. You would still have to admit that we have *means of acquiring beliefs*. We have sensory organs, and we have brains that process the information from those sensory organs. Our eyes, at least, are surely inbuilt in the sense of being highly likely to develop in the same way in many different environments, and products of evolution. Evolutionary psychologists have also argued that many of the specific ways in which our brains process information are inbuilt as well. Once again, I shall leave these arguments for Chapter 13. At any rate, the claim about our sense organs themselves is hard to gainsay. We now have our second conclusion.

(ii) In so far as we have innate (etc.) means of acquiring beliefs, they are highly likely to be reliable, as they are highly likely to be products of natural selection.

Some such claim is needed if we want to use the evolutionary argument to guarantee the truth of many of our beliefs about things in the modern world, as opposed to our beliefs about lions, zebras and the like. Our world is filled with things that our distant ancestors had no dealings with, such as cars, telephones and bank accounts. Natural selection has not weeded out people who have false beliefs about these things. Reliable means of acquiring beliefs, however, are presumably not confined to things our ancestors encountered.

Quine's thought is similar to that illustrated by the example from Hofstadter. As Quine was well aware, the objects of our experience can be carved up in different ways. To do induction at all, you have to carve the world up in *some* way. Quine's thought is that natural selection has seen to it that we generally carve things up in the *right* ways.

So natural selection seems to provide us with a guarantee that our beliefs are highly likely to be true. In some respects, this view is more like coherentism than foundationalism. Foundationalism usually picks out a special class of beliefs as the ones that have certainty. Beliefs arising directly out of sense-data are the most usual candidate. This evolutionary reliabilist epistemology is not limited in this way. The argument for our beliefs being reliable can be applied to *any* kind of belief, provided that that belief, or beliefs acquired in the same way, made a difference to our distant ancestors' evolutionary fitness and, it should be added, provided that the right beliefs are physically attainable, within the creature's cognitive capacities and so on. (This last-mentioned factor will be discussed again shortly.) This is not confined to things our ancestors could see. Recall Darwin's argument that, unlike artificial selection, natural selection can act on the innermost secret parts of organisms, provided those parts make a difference to their survival and/or reproduction. Similarly, natural selection can act on beliefs, or means of acquiring beliefs, about the innermost secret parts of the world, provided they make a difference to the believer's survival and reproduction. For example, it is often claimed that we have evolved beliefs, or at least the means of acquiring beliefs, about the beliefs and desires of other people. Yet you cannot *see* the mental states of other people. Foundationalist epistemologies have often run aground on the question: how do we know other people have minds? Evolutionary reliabilist epistemology, it seems, dissolves such worries, because it is useful to know what other people are thinking.

Note that the natural selection argument only leads to the conclusion that our beliefs in general are *highly likely* to be true, not that any of them is guaranteed to be true. It means that most, not necessarily all, of our beliefs are true. A similar claim is made by coherentists. But in the case of the coherentists, the reason for accepting this claim is a rather arcane, and contentious, philosophical argument that allegedly shows that a set of beliefs that are not mostly true is a conceptual impossibility. By contrast, the evolutionary reliabilist claim rests on a down-to-earth argument about whether people survive or not. But the evolutionary reliabilist can say, with the coherentist, that we can test a hypothesis by comparing it with our total set of already existing beliefs. If it does not fit in with many of our already existing beliefs, then it is probably false.

However, this is not the only way we have of testing it. The evolutionary reliabilist can give a special importance to information from the senses, since our sense organs have evolved to be reliable. This is a slight move towards the foundationalist side, but there are still significant differences. The evolutionary reliabilist view says that our senses are highly reliable about the world around us. The foundationalist view says that our senses are infallible about the fact that they tell us what they tell us. That is, the evolutionary reliabilist says that your senses tell you, reliably, that there is a tree in the quad. The foundationalist says that your senses tell you, infallibly, that it seems to you that there is a tree in the quad. The foundationalist then has to give an account of how we rationally *infer* from this that there is a tree in the quad, but it is not quite clear how this works. The evolutionary reliabilist says we do not have to worry about this, because our sense organs were shaped by natural selection to give us reliable information.

11.3 The limitations of our minds

Things are not quite as simple as this optimistic picture suggests, however. The account of how some beliefs get weeded out and some flourish depends on the plausibility of the assumption that the ones that get weeded out are (by and large) the false ones, and the ones that flourish are the true ones. But strictly speaking, this being an adaptive explanation, beliefs flourish because they are *beneficial* (spelt out in some suitable way), rather than because they are true. In some ways this is similar to one of the objections to Popper above. Popper wanted to characterize the development of science as progress, understood as getting closer to the truth. But all that the natural-selection analogy seems to warrant is a story of theories being accepted or rejected for various reasons. Ultimately, the ones that are accepted are the ones that are "adapted" to their environment, not necessarily the ones that are true. Similarly, the evolutionary reliabilist wants to characterize our beliefs as guaranteed to be mostly true, or our senses as reliable, in the sense of likely to generate true beliefs, in virtue of the power of natural selection. But all that the natural selection analogy warrants is seeing our beliefs and senses as having been selected in virtue of their contribution to fitness for various reasons. Ultimately, those that are selected are selected because they *were* useful in our ancestors' environment, not necessarily because they are true or, in the case of the means of acquiring beliefs, likely to tell us the truth.

So we cannot, seemingly, use the evolutionary reliabilist argument as it stands to show that we have inbuilt beliefs that are likely to be true. At best it shows that we have inbuilt beliefs that are likely to be useful. Worse still, they are not even guaranteed to be likely to be useful *to us*, but only to have been useful to our distant ancestors. Similarly, whatever means of acquiring beliefs evolution has bequeathed to us, they are only guaranteed to have produced beliefs that were useful to our ancestors. There is a still further complication here, as things are not necessarily selected in virtue of their usefulness to an organism, but in virtue of their usefulness to its genes or some other replicating entity. However, that need not be a major problem because what benefits an organism will usually benefit its genes and vice versa. So I shall disregard this complication here.

How do we deal with the fact that it is our ancestors, not us, that beliefs and means of acquiring them are selected to benefit? Clearly, the problems we face are not the same as those our ancestors faced. We have to cope with things that our ancestors did not, for example, air travel, escalators and other hazards. Moreover, human beings live in a wide range of different natural environments, much wider, probably, than those of our Stone Age ancestors. How could natural selection have prepared us for them? Two answers can be given here.

First, although our environment is different it is not *utterly* different. Solid objects still cannot pass through other solid objects, various fruits are still edible, animals still move without having to be pushed and so on. This point is quite strong, especially if we bear in mind that we have inherited *means of acquiring beliefs* rather than just beliefs. Suppose our eyes and the visual-processing systems in our brains are the same as those of our Stone Age ancestors. Surely, they were designed by natural selection to process visual information taking into account the principles of optics. For example, an object that is actually uniformly coloured and does not change colour does not reflect light uniformly, and moreover reflects different amounts of light in different lighting conditions. Yet our eyes are designed to cope with this; we are not normally fooled into thinking something has changed from one time of the day to another. But those laws of optics have not changed since the Stone Age, so if our ancestors' eyes informed them reliably, then our eyes inform us reliably too.

Secondly, paradoxically, we can make the reliabilist argument stronger if we change the scenario from a Stone Age one where, as in the stories about cave people above, getting things right was a matter of life or death, to the milder conditions of today. For it is clear that, however different our world is from the Stone Age one, we manage to negotiate it successfully. That is, we do not generally walk into walls, eat firelighters or jump off moving trains, or at least if we do we know what we are doing. Note that this cuts right across the distinction between what was around in the Stone Age and what was not. Note, also, that not all the hazards we successfully manage to avoid are life-threatening. We not only survive but manage to negotiate the world pretty comfortably, which surely shows that we are getting things right. This argument does not rely on any appeal to our being products of natural selection: a biblical literalist six-day creationist could accept it. But it does share a common core with the evolutionary reliabilist argument. On both views, the fact that our beliefs are helpful, evidenced by the fact that we do not come to grief, shows that they are likely to be true, or that our means of acquiring them are reliable. We could call such an argument "reliabilist" without the "evolutionary" bit.

Parallel arguments can be found in philosophy of science. One of the most popular arguments for scientific realism is the "no miracles" argument (see Putnam 1975). Very briefly, the argument goes: ask yourself the question "Why do scientific theories, and mathematics, achieve the empirical and technological successes that they do?" If they do not capture genuine features of the world, it is argued, then these successes would be miraculous. The best explanation for the successes, then, is that the theories capture genuine features of the world. Further, we can look at the history of science and see it as a story of progressive improvement, in that our ability to predict and control is progressively increasing. This is to be explained, according to the realist, by the theories being progressively better and better approximations to reality.

Whatever the merits of these arguments (and there have been plenty of rejoinders to the "no miracles" argument), they do not dispel the worry I mentioned above that what is guaranteed is usefulness, not necessarily truth. We were told that the fact that we negotiate the world pretty comfortably shows that we are *getting things right*, and that science captures genuine features of the world, and were to conclude from that that our beliefs, and our science, are largely true. But this is quite a step. Surely, we want to say, we cannot equate what is true with what it is useful to believe. There is a school of philosophers who do just that: the American pragmatists James and Dewey, and in our day Rorty. But, unless you make usefulness of belief your definition of truth, it seems clear that something can be useful to believe without being true, and vice versa. James Joyce's birthday was 2 February. That is true, but it is not particularly useful for me, or most people, to believe it. Moreover, pragmatist accounts of truth are in danger of slipping into relativism. It might be useful to know Joyce's birthday if you are writing his biography; is it then true for Richard Ellmann, but not for most people?

Nietzsche suggested that beliefs that are false can often be more useful than those that are true, especially those about the fundamental things in life. Believing in God may be useful in all kinds of ways, such as promoting social cohesion, deterring potential criminals and the like. But the very naturalists who try to guarantee the truth of our beliefs by appeal to evolution would not say that *just for that reason* it is true. (Nor would most believers in God.) One still might say, however, that our beliefs about the everyday world around us are likely to be useful if and only if they are true. If there is a lion in the distance, it is better to know that there is. There does not seem to be any plausible story about how believing the opposite could promote social cohesion and the like that could override this. Likewise, it is useful to have sense organs that tell you the truth about how far away things are, how big they are and so on. It may not be entirely clear where this leaves science, because science often tells us things that flatly contradict our everyday experience. For example, science tells us that inanimate objects do not have to be pushed to keep on moving, and that is extremely mild compared to the weirdness of quantum mechanics. But at least science is *answerable* to the beliefs that we acquire by means of our senses. Presumably evolution gave us the capacity to revise our beliefs in the light of thinking about what our senses tell us.

But how much elasticity can there be in this ability to revise our beliefs? Although this move might cheer us up a little about our everyday beliefs, it seems to leave us vulnerable to what is known as *cognitive closure*. Brains that were designed by natural selection to get around in the Stone Age may come to be equipped with mechanisms for reliably getting at the details of the physical world. These mechanisms may have enough elasticity to think usefully, and even arrive at the truth, about things well outside those Stone Age ancestors' experiences. But there must surely be some limitations to this. Colin McGinn has suggested that the mind–body problem is one such limitation.

> The problem arises, I suggest, because we are cut off by our very cognitive constitution from achieving a conception of that natural property of the brain (or of consciousness) that accounts for the psychophysical link. This is a kind of causal nexus that we are precluded from ever understanding, given the way we have to form concepts and develop our theories. No wonder we find the problem so difficult! (1991: 2–3)

You may be surprised to learn that the evolutionary psychologist Pinker (1997) agrees with McGinn on this. That may seem an odd attitude for a hard-nosed scientist, particularly one who is in close league with Dennett, who strongly disagrees with McGinn on this issue. But, as we shall see in Chapter 13, it arises from evolutionary psychologists' commitment to the *modularity thesis*: the thesis that the mind is a collection of tricks evolved to solve specific problems, rather than an all-purpose reasoning machine. The wonder, says Steven Pinker, is not that there are things forever beyond our understanding, but that we can understand as much as we do. Since the modularity thesis is justified, in part, by evolutionary arguments, it seems that the theory of evolution lends strong support to the claim that there are strict limits to what we can understand. One might feel a little short-changed by this conclusion, however. Surely we already knew that our minds are limited? However, a greater knowledge of our fundamental psychology, guided by evolutionary thinking, might give us some clues as to what the specific contours of our limitations are.

12. Evolution and religion

The theory of evolution is often seen as a rival to the theory that the world was created by God. Popular media sources, including numerous websites, often either point out alleged flaws in the theory of evolution with a view to promoting belief in God, or attack a view called "creationism" in the name of good science, and in particular evolution. In these latter attacks it is not always clear exactly what creationism is, but in the crudest versions the word often means a literal belief in the account of creation in *Genesis*. But it is clear that many religious believers, including many members of mainstream Christian churches, neither believe the *Genesis* account literally, nor are required by their churches to do so. The Roman Catholic Church, for example, has always taken the view that the Bible should not be read literally.

There are, of course, more sophisticated versions of the anti-creationist argument that do not focus on an account as specific as that of *Genesis*, but instead on the broader claim that the world was created by a God as standardly conceived by the monotheistic religions: a being that is omnipotent, omniscient, omnibenevolent and so on. The purposefulness of many features of organisms (the eye being "well designed" for seeing and so on) has frequently been taken to be especially compelling evidence for this God. So the theory of evolution, offering as it does an alternative account of why these features exist in terms of purely non-goal-directed forces, is often seen as reducing the force of this evidence.

In one way or another, it seems to be a widely held view that the theory of evolution lends support to atheism. I shall examine some of these ways in the first section.

12.1 Does the theory of evolution support atheism?

Many religious believers are content to say that there is no conflict between accepting the theory of evolution and believing in God. However, a number of arguments have been advanced against this. It is widely held in both academic and popular circles that anyone who accepts the theory of evolution has reason for believing that there is not a God. Why might people think this?

12.1.1 Is theism rendered superfluous as an explanation?

One reason might be that the two hypotheses are *rival explanations*. That is to say, whereas the apparent designedness of living things was once explained by the existence of God, it is now explained by evolution. Thus, God is rendered a superfluous hypothesis. (Note that this argument assumes that evolution is a *better* explanation, but since the claim I am presently addressing is that *if* we accept evolution *then* we can do without God, that need not detain us.)

There are a number of possible replies to this, but as it stands it can be dealt with by means of one simple consideration. That is, the above argument is clearly a *non sequitur*. It relies on the premise that the designedness of livings things is the only thing that God might be invoked to explain. This is clearly false, since God is generally held by believers to be the creator of the entire universe, not just the living things in it. A fallback position that has often been adopted is to argue that whatever other things God might explain can be explained by evolution also. (There is, of course, the possibility that there could be other explanations for the other things that involve neither evolution nor God. But the matter at hand is the oft-claimed relevance of *evolution* to the question of God's existence.) Theists often argue that, while evolution explains the existence of living things, there are other things that it does not explain. But what are these "other things" that need to be explained?

Initial conditions

One candidate can be proposed as follows: evolution explains (or is purported to explain) how life came about *given that the physical world is the way it is*. Or, more precisely, given the way it was at the time of the life's origin, since living things have significantly altered their surroundings (free oxygen in the earth's atmosphere due to the photosynthesis of plants, existence of soil due to earthworms, and so on). But does evolution explain how those initial conditions themselves came about? If not, then its explanation for how life came about is at best incomplete.

This case can be bolstered by giving instances of the relevant conditions. I will let one example stand for the many that have been suggested.[1] The existence of living things is only possible if highly complex chemical compounds can be formed. This in turn requires the existence of a chemical element whose atoms can bond with a large number of other atoms at the same time – to put it more technically, an element with a high valency – and there must be a fair amount of this element in one place, at that. This condition is fulfilled on earth by the presence of carbon, which has a valency of 4. (In science-fiction type speculations, it has often been suggested that silicon-based life might exist on other planets, as silicon also has a valency of 4.) Carbon is produced in the interior of stars. Stars initially burn hydrogen atoms to form helium atoms. When this has gone on long enough, the star contracts and the interior gets hotter until the helium ignites. This sets up a chain reaction, whereby helium atoms fuse to form beryllium atoms, which then fuse with more helium atoms to form carbon. Moreover, oxygen atoms are formed by a further step: carbon atoms fusing with yet more helium atoms. Both carbon and oxygen are crucial for life, but a slight adjustment to any of a number of physical constants would have made it impossible for these two elements to be formed. If carbon were produced at a faster rate, there would be very little helium left to produce oxygen. Conversely, if it were slower, most of it would be burnt into oxygen due to the presence of large amounts of helium. Hence, the argument goes, a delicately balanced set of conditions, necessary for the coming into existence of life, was in fact met, and this is a state of affairs that is intrinsically improbable and cannot be explained by evolution.

Moreover, the conditions under which life came about also include the fact that the universe is governed by *this* set of laws rather than some other set. Alternative possible sets of laws might not have allowed atoms to exist at all, let alone atoms of high valency. Again, the argument runs, we have an intrinsically improbable state of affairs that is necessary for the formation of life; it is actually met, and evolution cannot explain it.

The second argument, however, undermines the first one, and moreover is not as strong as it might at first appear. The first argument says that high-valency elements would not have been formed if the rate of formation of carbon had been different. But that is only the case *provided all the other physical laws were the same as the ones we actually have.* For all the argument says, there could be any number of other possible sets of laws under which the formation of high-valency elements (or some analogue) would have been much *more* likely. Moreover, the second argument only works if we can establish that the set of possible sets of laws that would allow life to come into existence is much smaller than the total set of possible laws. But how do we know this? There are computer-generated simulations of evolution that are based on much simpler "laws of physics" than the actual ones.[2] So we cannot say that *only* this set of laws makes life possible. Do we have a clear idea of just how broad or narrow the set of possible sets of laws is that would make life possible? Perhaps there are many alternative sets of laws under which the coming into existence of life would have been *much more likely* than under ours. In the absence of clear answers to these questions, the argument must be considered inconclusive.

Laws

Let us say that it is allowed that life could have come into existence in a world with different initial conditions, or even with different laws, from the actual world. Still, it might be asked, could life have come into existence in a world that had *no laws at all*? Since evolution works by entities replicating themselves, the conditions in which those entities live has to be stable and predictable in a whole host of respects. This is partly because lineages of organisms adapt to their environment by a process of gradual accumulative change, which would not be possible were there not conditions that were stable and predictable for long enough for adaptive modification to take place. Moreover, any organism keeps itself alive by constantly employing resources from its surroundings (food and air, to name the most obvious examples in our own case). If there was not stability and predictability regarding these resources, the maintenance of such a process, not to mention the replication of the same pattern – that is, life cycle – over generations would be inconceivable. As Hume writes (although making a different point):

> The bread, which I formerly eat, nourished me; that is, a body of such sensible qualities was, at that time, endued with such secret powers: but does it follow, that other bread must also nourish me at another time, and that like sensible qualities must always be attended with like secret powers? ([1748] 1999: IV, pt II)

Hume's point (or part of it) is that there is no logical necessity that the bread that nourished me in the past will nourish me in the future. But if we, and living things in general, could not rely on such regularities, life would be impossible. Moreover again, in a world without laws, even the basic microchemical processes that underlie life – metabolism, replication of DNA and so on – could not occur. It is not just that *those* particular microchemical processes require *this* set of laws (something that is trivially true anyway). It is that regularly recurring processes are needed for there to be any kind of life, and these processes would be impossible were there no laws at a lower level on which they could be sustained.[3] But, as far as we can tell, the fact that the universe has laws at all is *contingent*: that is, it need not have been the case. It appears, then, that we have after all a state of affairs that is necessary for the formation of life, that is actually met and that evolution cannot explain. We may as well add intrinsically improbable as well, since surely there are far more possible scenarios that are disordered than ones that are ordered.

To put this another way, evolution at best explains how life originated, but there are conditions that need to obtain for the process described for the theory to get going at all. These conditions cannot be explained by evolution. They lie, so to speak, outside its remit. Since God is supposed to be an explanation for why the world in general is the way it is, not just for life, the theory of evolution cannot, by itself, be a rival to theism.

Yet this argument may rely too heavily on a verbal quibble abut the word "evolution". It is true that the theory of evolution is a biological theory, and therefore cannot be expected to explain, still less to be a rival to theism as an explanation for, anything non-biological. But perhaps some process analogous to evolution can explain the troublesome non-biological facts mentioned above. If the analogy is a good one, then we could just as well call the process "evolution", and say that evolution does supply an alternative to theism as an explanation even for these non-biological facts.

Attempts have been made to construct a model for explaining the existence of a law-governed world on lines that are explicitly intended to be analogous to natural selection. Hume made a proposal in *Dialogues Concerning Natural Religion* ([1779] 1998: pt 8) that has been acknowledged by advocates of the "evolution dispenses with the need for God" view as broadly analogous to a natural-selection type explanation (e.g. Dennett 1995: 32–3). Hume begins with the point that *of course* we live in a world where the laws are such as to enable the coming into existence of life, since we are in it.[4] According to Hume we can posit a completely undirected set of initial conditions that would allow such a world to come into existence by means of random variations. He proposes that there may have been a fecund random generator of worlds: that is, a mechanism in which worlds were constantly being generated at random, thus requiring no intrinsic design or directedness. These worlds could vary with respect to their laws just as they could vary with respect to such features as how much matter or how much heat they contain, and if worlds without laws are possible, then they would be included as well as ones with laws. On this scenario, if we wait long enough we are bound to get a world with laws that favour the creation of life. Trivially, we live in such a world. We call the laws of this world "*the* laws of physics". But we have no reason to think that there are not many, many other worlds out there. This dispenses with the need to estimate just how likely is the formation of a universe whose conditions allow the formation of life, since we can postulate a random world generator as fecund as we like.

Another scenario Hume suggests is a universe – just one is needed for this – containing a finite amount of matter randomly moving about over an infinite amount of time. If this were the case, then every possible combination of that matter – and every possible sequence of combinations of that matter – would eventually come into existence. This includes, of course, the combination or sequence that was required to produce us. It would also include sequences of combinations that behave in a law-like manner, since all sequences that are regular are possible sequences.[5] These two suggested scenarios have a common basic pattern: starting with a randomly generated profusion of types of universe – which exist either simultaneously or one after another – we will inevitably get one that allows life to come into existence.

Admittedly, Hume offered these suggestions among a host of other possibilities – such as the universe being like a great animal, or being spun as a spider spins a web – and his conclusion seems to have been that none of these possibilities was any more satisfactory than any other. However, it is these two, or at any rate the common idea underlying them, that has attracted the notice today of people who believe that evolution dispenses with the need for God. The analogy to natural selection is hopefully clear: the basic thought is that the fittingness of the parts of organisms to the functions those parts perform is explainable by purely non-directed forces, and ultimately by the trivial truth that if those parts did not perform those functions, the organisms would not exist. Similarly, the fittingness of the world to support life is alleged to be explainable by purely non-directed forces, and ultimately by the trivial truth that if the world could not support life, life would not exist.

However, these possible alternative explanations to theism for the way the world is are no more than that: *possible* alternatives. We have no evidence that either a random world generator or a finite-matter-plus-infinite-time scenario exists. The reason that the biological theory of evolution is so appealing may, for some people, be its success in eliminating teleology. But we have considerable evidence for accepting the theory of evolution beyond that. Among other things we have evidence from the fossil record that a succession of different creatures existed, and evidence that creatures produce offspring with variations. By contrast, we have no evidence that other universes exist, or that random new types of universe are generated.

It is often suggested (and again Hume gave hints of this long before Darwin) that an explanation involving only non-directed forces of some kind is *simpler* than one involving God, since God is conceived as a being with a mind and therefore highly complex. Thus, this argument goes, we should reject God as an explanation by Ockham's razor. However, it is not clear that we should think of God as complex. Richard Swinburne (1991) makes a case that we should not, arguing, for example, that the features of omnipotence and omniscience make God considerably simpler than a human being. It is not clear which way Ockham's razor cuts here. So it seems that satisfying a desire that some people have to eliminate teleology is the *only* justification for accepting either of these scenarios. This in turn seems to depend on a prior allegiance to atheism, and is not justified by any empirical evidence. To eliminate theism in favour of one of these alternatives would be no more than speculation. So once again the argument is inconclusive.

12.1.2 The problem of evil

It is sometimes suggested that the (misleadingly named) problem of evil – that is, the difficulty of reconciling the suffering in the world with a God who is both omnipotent and omnibenevolent – is exacerbated by the cruelty and inefficiency of natural selection. Darwin himself made this connection in a remark that has become famous: "What a book a devil's chaplain might write on the clumsy, wasteful, blundering, low, and horribly cruel works of nature!" (letter to J. D. Hooker, 13 July 1856, in Darwin & Seward 1903: vol. 2, 94). He briefly returned to this theme at the very end of the chapter in the *Origin* entitled "Struggle for Existence". In an apparent attempt to play down the harshness of the picture he has been painting, he writes: "When we reflect on this struggle, we may console ourselves with the full belief, that the war of nature is not incessant, that no fear is felt, that death is generally prompt, and that the vigorous, the healthy, and the happy survive and multiply" ([1859] 1968: 129).

If this is meant to be a reply to the "devil's chaplain", it is clearly an inadequate one. First, it presupposes that animals do not feel fear – not to mention pain – which is highly questionable. Secondly, the claim that "death is generally prompt" can be answered with a host of counter-examples. Thirdly, it does not at all address the issue of *inefficiency*, which is over and above any cruelty. That Darwin was aware of this issue is indicated by the words "clumsy, wasteful, blundering" in the quote above. On the model he himself proposes, natural selection requires the production of many more organisms than can reproduce, and in reality, as he points out, this means that many do not reach the age when they can reproduce. This does not sit well with the idea of an omnibenevolent deity, as that is presumably supposed to mean a deity who is concerned for the wellbeing of *all* creatures. A great many creatures, on a Darwinian scenario, have no role in the cosmic drama other than to be born and then die very soon afterwards. I shall return to the "inefficiency" issue at the end of the next section, but I shall say a little here about the "cruelty" issue.

The theory of evolution may lead us to revise upwards our estimate of the amount of suffering there is in the world. But this is not a fundamentally new consideration. The perception of there being an overwhelming amount of suffering in the world was readily available long before Darwin, and numerous quotes from Sophocles, Shakespeare and many others could be produced to show this.[6] I do not propose to rehearse the time-honoured replies to the problem of evil here. But if those answers were adequate in the face of earlier perceptions of the overwhelming misery of the world, then they should remain adequate in the face of the horrors unveiled by Darwin. If they are inadequate after Darwin's unveiling, then they were inadequate before. A similar point was eloquently expressed by G. K. Chesterton:

> The materialism of things is on the face of things; it does not require any science to find it out. A man who has lived and loved falls down dead

and the worms eat him. ... If mankind has believed in spite of that, it can believe in spite of anything. But why our human lot is made any more hopeless because we know the names of all the worms who eat him, or the names of all the parts of him that they eat, is to a thoughtful mind somewhat difficult to discover. ([1908] 2004: 79–80)

The theory of evolution at best (or worst) alerts us to some new and vivid examples of suffering, but it adds nothing fundamentally new to the traditional problem of evil, with the possible exception of introducing the idea of the "inefficiency" of the natural-selection process. Many answers to the problem of suffering already exist in religious literature. If these answers are tenable, evolution adds no fundamentally new considerations that make them any less tenable.

12.2 "God of the gaps" arguments

The arguments for theism that I discussed in §12.1.1 are part of a more general class of arguments known as "God of the gaps" arguments. The general form of these arguments is: science cannot explain X; God can explain X; therefore X is evidence for the existence of God. Clearly, science includes much more than the theory of evolution. Even if we extend the term "evolution" to include Humean analogues to evolution discussed above, there are clearly many other types of scientific explanation. However, the theory of evolution is the only naturalistic explanation we have for why there are things that give the appearance of having been designed for a *purpose*. The theistic arguments that appeal to the alleged improbability of there being a universe capable of supporting life, are implicitly arguing that such a universe gives the appearance of having been designed for a purpose. It is no coincidence, then, that quasi-evolutionary models have been proposed by committed naturalists to explain away that appearance.

It might be argued that we need God to explain why there is a universe at all, and that evolutionary explanations of any kind would run out here. There would surely need to be *something* in existence for a quasi-evolutionary process to get going. But, *prima facie*, a universe completely devoid of life, and incapable of producing life, would not bear the appearance of having been designed for anything. So it would not call for explanation *either* by God or by some quasi-evolutionary process.

In the first section of this chapter, I said that I was taking for granted that evolution was capable of explaining how life came about, given the initial conditions at the time it originated. However, this has been challenged, and not just by the scientifically illiterate, but by a vocal minority of reputable life-scientists. One of the most highly publicized challenges comes from Michael Behe (1996), a professor of biological sciences. The phenomenon that he considers evolution incapable of explaining is what he terms "irreducible complexity". He defines an irreducibly complex system as: "a single system which is composed of several well-matched, interacting parts that contribute to the basic function, and where the removal of any one of the parts causes the system to cease functioning" (*ibid.*: 39). The thought is this: imagine an apparatus that consists of a number of interacting parts. If one of those parts were not there, the apparatus as a whole would not work. As an example, he asks us to consider a mousetrap:

> A common mousetrap has several parts, including a wooden platform, a spring with extended ends, a hammer, holding bar and catch. Now, if the mousetrap is missing the spring, or hammer, or platform, it doesn't catch mice half as well as it used to, or a quarter as well. It simply doesn't catch mice at all. (*Ibid.*: 280)

Let us leave aside the problems we saw in Chapter 7 around defining "function". And let us forgive Behe for the fact that nothing would, strictly speaking, fit his definition of "irreducible complexity", since the removal of one of its parts would only cause the system to cease functioning if something else was not put in to replace that part. For the sake of argument, we can agree that a mousetrap clearly serves a purpose, and that it is clear what that purpose is. Likewise, in the case of many biological organs, we can clearly see that they have purposes and what those purposes are. Behe's point is that, in many cases, if we remove a part the apparatus no longer serves its purpose. The complexity is irreducible in the sense that carrying out the function depends on all the parts being there. Hence, Behe reasons, such apparatuses could not have come about by a gradual process such as evolution. For that to happen, there would have to be a succession of useless apparatuses that eventually led up to a useful one. Barring massive coincidence, then, we would have to conclude that the apparatus was designed.

According to Behe, we actually find many examples of irreducible complexity in microbiology. He devotes a chapter of his book to the blood-clotting system, and also discusses the flagella (hair-like structures used for locomotion) of single-celled organisms. Behe denies that his argument is a God of the gaps argument. That is because he refuses to say explicitly that his conclusion is that there is a God, instead only saying that there is "intelligent design". However, this denial is unconvincing. An intelligent agency that designed the fundamental building-blocks of life sounds very like God. Admittedly, it would not have to be the God of the Judaeo-Christian religions. It would not have to be all-powerful, just powerful enough to create life; and it would not have to be benevolent, to respond to prayers or provide a life after death, for example. Nonetheless, a very powerful intelligent designer could still reasonably be called "God". Let us not quibble over terms. Suffice to say that most atheists and naturalists would deny the existence of Behe's intelligent designer. In any event, whatever he may say himself, Behe's argument has been seen by many theists as evidence for the existence of God (e.g. Plantinga 1991), and George W. Bush's favourable remarks on "intelligent design" may well have a Beheian inspiration.

Unsurprisingly, Behe's argument has been attacked by atheists on a number of counts. It has been pointed out that it is a *non sequitur* to say: without all its parts *X* would not do what it does, therefore without all its parts *X* would be useless. Behe is accused of ignoring the possibility that the remaining parts could do something else. Kenneth Miller (2003) has specifically attacked the mousetrap example, arguing that even if you took away one, or two, or three parts, it would still be useful for something. For example, an incomplete mouse-trap could be used as a tie-clip, a catapult or a fish-hook. As far as biology goes, we know of many cases where an organ that does one thing evolved out of one that does another thing. The human hand is a modification of a foot; lungs are modified swim-bladders. Miller (2004) also offers a counter-example from the world of bacteria to Behe's claim about the flagellum. Some parasitic bacteria have a type three secretory system (TTSS), which uses some of the same proteins as are in flagella to inject chemicals into their hosts. The TTSS is, so to speak, an "incomplete" flagellum, but it is still useful for something.

Leaving aside these empirical objections to Behe's argument, two more general points can be made. First, there seems to be an inconsistency in Behe's position. He holds that once single cells had been designed and built, evolution could proceed as Darwin's theory describes, and produce all the multi-celled organisms we see around us. However, surely if his concept of irreducible complexity applies to single cells, it applies to the macroscopic organization of multi-cellular organisms as well. For example, your heart and your lungs would not serve the purposes they do without each other. Your lungs oxygen-ate blood, but they need blood to come to them before they can do that, and oxygenated blood would be no use to the organism if it were not circulated around the body. On the other side, the heart pumps blood around the body, but that would be of no use to the organism if the blood were not oxygenated. And there is much more "irreducible complexity" here: the diaphragm muscle is needed for your muscles to breathe; the veins and arteries are needed for blood to circulate through; the kidneys are needed to remove waste products from the blood; and so on. If there is no problem explaining this state of affairs, it is not clear why there is supposed to be a problem explaining the "irreducible complexity" in single cells.

Admittedly, this is an *ad hominem* criticism. Anti-Darwinians might go further (and some have) and say that the irreducible complexity at the macro-level cannot be explained by evolution either. But then they are up against the fact that we do actually have pretty good knowledge of the evolutionary history of some organs, showing that they did serve purposes in simpler forms than they are in now.

A still more general point about God of the gaps arguments, including Behe's, is that they are at the mercy of future science. The most these arguments can ever entitle us to conclude is that we have not come up with an explanation for something *yet*. The history of science is a history of people solving what seemed to be intractable problems, such as the precession of the planets or how bumblebees are able to fly. One thing we can be reasonably sure of is that science will eventually solve at least some problems that now appear intractable. We do not have to believe that science will one day explain everything, but the fact that science is currently unable to explain something does not show that it never will.

So the upshot so far looks rather inconclusive. I have argued that accepting the theory of evolution does not give one a reason to deny the existence of God. I have also argued that limitations on what evolution (or science in general) can explain do not give us a reason to believe in God. Belief in God and belief in evolution are, in the view of a great many religious believers, perfectly compatible. "Compatible" here can mean a little more than the weak sense of logically non-contradictory. Evolution provides a general template for explaining how life came about in a world of purely non-goal-directed natural laws. It allows us to explain how life came about without having to posit any processes over and above those described by those natural laws. It does not require special acts of intervention in those laws to create life. But a universe governed by natural laws that are the same everywhere is, it can plausibly be claimed, exactly what one would expect God to create. Surely, a being that was omniscient and omnipotent would be capable of setting up the laws of physics so that living things would come about by the operation of those laws alone. The alternative, which seems to be favoured by the intelligent design hypothesis, seems to be that God needed to do further work specifically to bring about life. Why could God not have known beforehand what laws would allow life to evolve, and put those laws into place?

Swinburne (1991: ch. 12) has argued that a world with a set of universal laws is *better* than a world where special acts of intervention take place. One reason is that such a world is more intelligible and hence easier to negotiate. How would we get around in a world in which laws of nature were occasionally violated? We can also add that the intellectual satisfactions and the cultivation of intellectual virtues that are involved in the practice of science are made possible by the regularity of the world. Swinburne uses this as an answer to the problem of evil. It is the price we pay for living in a world governed by natural law. As Dawkins has argued, natural selection is the only process we know of that could produce life out of the laws of physics. If it is in fact the only way *tout court*, then a God who wanted to create life in a law-governed universe would do so by means of natural selection. It appears that natural selection cannot but be a cruel and wasteful process, but that is the price of having life in a law-governed world at all. If this is right, then the accusation of "inefficiency" loses its force.

I shall add one final consideration here. A world in which life came about through the operation of natural laws that were set up by God, rather than through special intervention, would to us look indistinguishable from a world in which life came about in accordance with natural laws that are just brute facts. It does not need knowledge of modern science to appreciate this: as Chesterton said, the materialism of things is on the face of things. A universe in which God's actions are from our point of view indistinguishable from the actions of mindless nature allows an *unforced faith*. And many theists (e.g. William James) consider that it is better to have a choice about whether to believe than to be compelled to believe. This may annoy atheists, who may protest that such moves render belief in God unfalsifiable. However, we can treat this as an argument only for the compatibility of theism with evolution, not for the existence of God. Moreover, in any event, even Popper admitted that a theory can be unfalsifiable and yet be true. I shall say a little more about this reconciliation strategy at the end of the chapter.

12.3 Evolution and explaining religion

In §12.1 we looked at some attempts to show that evolution renders religion redundant as an explanation. These arguments all, in one way or another, entail the thought that religion once seemed to provide the best, or only, explanation, for some phenomenon for which evolution now provides an alternative, and better, explanation. There is yet another such phenomenon where evolution has been claimed to provide a better explanation, and that is the phenomenon of religion itself. There is a long tradition of non-believers who offered a story about how religions originated. Notable works in this tradition include Hume's *The Natural History of Religion* and Freud's *Totem and Taboo*. Religious apologists sometimes challenge non-believers with questions of the following type: if there is no God (or gods, or spirits, etc.), then why have so many people believed that there is? The list of believers includes people at many different periods of history, and includes highly intelligent and well-informed people. The thrust of this challenge is that if we are unable to account for this fact naturalistically, we are forced to accept that those beliefs came about through the influence of supernatural agencies. This argument is not always used to support truth-claims about the doctrines of one particular religion; sometimes it is just used to support the claim that whatever engenders religious belief, it is something that is somehow beyond the scope of naturalistic explanation and understanding, and is therefore worthy of being the object of religious awe. This line is taken, for example in John Hick's *God and the Universe of Faiths* (1973) and *An Interpretation of Religion* (1989).

This can be seen as a variant of the age-old "argument from common consent", but in fact it goes beyond that argument. The argument from common consent says that the fact of widespread agreement *alone* furnishes us with good reason for believing in the truth of religious doctrines. But this argument says that the fact of widespread agreement, *which cannot be explained naturalistically*, furnishes us with good reason for believing them. Works that offer naturalistic explanations for the origin (and persistence) of religion seem well suited to meet this challenge. Such works, if they succeed, weaken the case for holding religious beliefs.

One might think: but what difference does an explanation for why people have religious beliefs make to whether or not those beliefs are true? In philosophy, to attack someone's beliefs on the grounds that they hold those beliefs for such-and-such a reason is called the *genetic fallacy*. (The word "genetic" here has nothing to do with genes, but both words come from the Latin *genesis*, meaning "origin".) An example of the genetic fallacy might be: the president says that the practice of constitutional review is a good thing, but she has a vested interest in this, because without constitutional review her power would be greatly reduced. We might conclude: therefore we can disregard all her arguments. But most philosophers would regard this as a fallacy. This is because the president has presumably offered *arguments* for what she has said; those arguments may be good or bad ones, but it should be possible to decide whether they are good or bad without considering whether she has a vested interest. Moreover, even if she had *no* arguments for what she said, it should still be possible to decide whether to agree with it or not, without considering whether she has a vested interest. One just has to think about any arguments for or against constitutional review.

However, if we think that coming up with a naturalistic explanation of religion somehow discredits religion, are we not committing the same fallacy? Should we not just look at the arguments for or against religion's claims? There are two reasons why things may not be quite so simple. First, a claim has been made that the fact that so many people are religious is somehow evidence for the truth of religion. This amounts to a challenge to non-believers to produce an explanation for why so many people believe that is consistent with those beliefs being false. Secondly, an explanation for something usually succeeds by being *better* than the alternative explanations. One of the alternative explanations for the ubiquity of religion is that religious beliefs are true. Non-believers will typically say that there is no evidence to support this. If they can produce an alternative explanation that does not have this disadvantage, then presumably that alternative explanation wins.

In any event, there are other reasons why one might be interested in giving a naturalistic explanation of religion. There may be a *therapeutic* aspect to such projects. Many of the works that seek to explain religion naturalistically also aim to persuade people to abandon religion. They can point to the irrationality of the grounds on which people hold religious beliefs, and couple that with the general inadvisability of believing things on irrational grounds, to furnish a persuasive argument for abandoning religion. The Hume and Freud works mentioned above seem pretty clearly to have therapeutic aims of this kind. Moreover, there is scientific and historical interest in the question in its own right. If you are a non-believer, then the official religious stories about how people came to have religious beliefs – that is, stories involving a divine revelation of some kind – will be unacceptable to you. But religion is such a widespread phenomenon, and has such a major influence on many people's lives, that there must be an interesting story (or stories) to tell about how it originated.

There are a great many mildly plausible stories that one could tell about why religion is so widespread. I shall briefly discuss a few before moving on to specifically evolutionary ones. One plausible explanation is that religious beliefs are popular because they are comforting: it is nice to believe that death is not the end, that an all-powerful being cares about you, and so on. But there are a number of problems with this. First, a belief can only make you feel better in this way if you actually believe it. It would be nice for me to believe that I am the most handsome man in the world, but I still do not believe it. Why do we believe some things that make us feel better but not others? Secondly, as Pascal Boyer (2001) points out, anthropological studies clearly show that a great many religious beliefs are not comforting at all. Many peoples around the world believe in witches with vast power and great malevolence. Even Christian beliefs, which seem to be what people usually have in mind with this explanation, are not necessarily all that comforting. The belief that God is watching you all the time can induce great guilt, as is well documented, for example in numerous books by Irish ex-Catholics (e.g. James Joyce, Edna O'Brien). Moreover, many Christian denominations make the chances of getting into heaven seem much less than the chances of getting into hell. The heaven–hell story can make life seem like an examination where the difference between the consequences of failing and passing is extremely great. But not many people find taking exams with high stakes a great comfort.

Another way one might try to explain religion is by the power of society, or of high-status individuals, over people's minds. People are highly susceptible to believing what is accepted in their society as a whole, and the word of an authoritative figure, such as a priest or a teacher, is likely to make an impression. This is all probably true as far as it goes, but it is seriously incomplete as an explanation. It leaves unanswered the question of *why* those beliefs are so prevalent in societies in the first place. Similarly, it leaves unanswered the question of why the authority figures say those things in the first place. A third approach might be to say that religion appeals to people because religion provides explanations for so many things. It explains why there is a world, why that world is structured, why the parts of organisms are so well designed, and so on. Dawkins (1986) argues that, before the theory of evolution came along, God provided the best explanation for why organisms are so well designed. However, Boyer (2001) argues that the demand for explanations for things is not as widespread as this suggests. Many people are happy to accept that things just are the way they are. For example, many peoples believe that the spirits of ancestors have great powers and make great demands, but although they often have detailed beliefs about what those demands are, and what will happen if they are not met, they equally often have no account of how ancestors are able to do the things they do. Moreover, any concept of God that can play such an explanatory role is quite a sophisticated concept, a God of theologians and philosophers. This is because, among other reasons, a non-believer can ask: what explains the existence of God? We might say, as many religious thinkers have, that God's existence is necessary. But the non-believer can reply: then why is the world's existence not necessary, and what does "necessary existence" mean anyway? And so on. The concept of necessary existence seems too sophisticated for the average religious believer. It seems unlikely that such a concept was involved in religion's origins. So we would need an explanation for why people think certain kinds of explanation are satisfactory.

Given that religion is so ubiquitous, it is likely that the explanation for why people are drawn to it involves things that lie deep in human nature. Chances are, then, that evolution will play a part in explaining it. One way in which it might do so is via meme theory (Dawkins 1993), which I introduced in §3.4. Religion is surely a very clear example of successful memes, in that it has been around for a long time and is very widespread. Moreover, it may not contribute to human survival or reproduction, in as much as people seem to be able to survive and reproduce just as well without it. Moreover again, it does not seem plausible to say that it was consciously designed to be a successful meme. For one thing, to say this would be to doubt the sincerity of religious believers. So religions seem to have some of the key features of memes: they replicate independently of the replication of individual organisms or genes, and they do so without having been consciously designed to do so. But remember that memetic explanations are supposed to be natural selection explanations, so memes – or at least collections of memes – should possess features that make them particularly well suited to replicating. Dawkins suggests that the doctrine of salvation and damnation is one such feature. In Christianity and Islam, this is usually coupled with the belief that one must have the right religious beliefs to be saved.

This seems a plausible explanation on the face of it: if one believes that one ought to do X or be damned, one has a strong motive to do X. But there is a problem with this. What is missing from the story is an account of why anybody would think that the whole faith/salvation/damnation story were true in the first place. Somebody might say to me, "If you take this pill you will live 10,000 years; if you do not you will die in five minutes. The pill costs five pounds". I would keep my five pounds, for the obvious reason that I have absolutely no reason to believe this story about the pill. But this is analogous to somebody telling me a story about faith and salvation and damnation. No matter how great the purported rewards and punishments, why would I believe the story? There must be something about human minds that makes people likely to find religious stories plausible or natural. To be fair to Dawkins, he is aware of this. His memetic explanations are natural-selection explanations and, as in any other natural-selection explanation, they involve a replicator being selected for its fitness in a particular environment. Whether a gene is successful depends on conditions in the environment it is thrown into, and the same must be true of memes. But memes "live" in human minds, so the way human minds are has an effect on whether a meme is successful or not. I shall return to this point shortly.

A second way that evolution might contribute might be to explain religion as some kind of by-product, rather than as an adaptation. That is, it may be that the human mind has certain features, that themselves have perfectly ordinary natural-selection explanations, that produce religion as a by-product. A suggestion along these lines is offered by Dennett (2006). He argues that religion may be a by-product of what he calls a *hyperactive agent-detection device*. The idea is that we have, as a product of natural selection, a capacity for detecting signs that an agent – a sentient creature – is in the vicinity. We can think of all kinds of reasons why such a device would be useful: it is useful to know if you are being stalked by a predator; it is useful to know if there are other human beings around, who might be potential friends, foes or mates. There is plenty of evidence that our visual systems are extra-sensitive to the presence of bilateral symmetry, which it is plausible to surmise is because, in the wild, bilateral symmetry usually indicates that something is looking at you. We also seem to be very quick to detect sentient motion as distinct from, say, leaves being blown by the wind. It is also useful to know if you have inadvertently walked into someone else's home, something that one could detect not by seeing the person but by seeing signs that they have been there, such as artefacts. So we can surmise that we have a capacity to detect signs of the work of a conscious agent. But, one might think, surely such a device is useful in so far as it is *accurate*. It is no use to us to detect signs of agency where there has not actually been any agency at all. This is where Dennett's argument takes a clever turn. He suggests that it is better that such a device detects *false positives* than *false negatives*. That is, it is better that it sometimes detects agency where there is none, than that it sometimes fails to detect agency where there is. This is because, Dennett suggests, the cost of a false positive is likely to be fairly small: an inconvenience, a little time wasted. But the cost of a false negative can be very high: loss of life in many cases. So our evolved agency-detection device is hyperactive, having an inbuilt bias towards detecting agents where there are none. This, then, Dennett suggests, can help to account for the origins of religion. A great many religions populate the world with agents: gods, spirits, witches, dead ancestors and so on. History suggests that more monotheistic, less anthropomorphic conceptions of God came along later.

Boyer's approach combines elements of memetics and evolutionary by-product explanation. He offers suggestions for why, and in what ways, the human mind is predisposed to find certain types of stories more plausible. For example, a large part of our cognitive architecture seems to be geared to negotiating social interactions. Given that human beings are very social creatures, and are generally very clever, we are likely to have evolved means of benefiting from cooperation, avoiding being exploited, detecting likely cooperators and exploiters, appearing to others to be a likely cooperator, figuring out what other people want from us and so on. Boyer points out that religions vary widely around the world, including ways that make the "comfort" explanation and the "explanation" explanation very unlikely. But what is extremely widespread, he argues, is belief in supernatural agents – gods, ancestors, witches and so on – who want something, and who have the power to help or harm human beings. These agents typically have unusual powers: they can see everything, they can move objects at a distance, and the like. But apart from that, Boyer argues, they tend to be agents with whom one deals as one would deal with another human being. For example, one wants to know what they want, and one accepts that people get upset when they do not get what they want. So, religious beliefs make use of a set of cognitive mechanisms that are already in place for evolutionary reasons.

But this, at best, helps to explain why religious beliefs *catch on*. It still does not explain how they originate. For that, we need to look to a memetic approach. Bear in mind that in any natural-selection explanation, the mutations can be generated completely randomly. Boyer argues that our minds randomly generate stories all the time. We constantly daydream, fantasize, have imaginary conversations and vaguely entertain thoughts. Sometimes we believe things for no reason, or for no rational reason anyway. But some of these stories, he suggests, catch on because they are stories of a kind that we find easy to understand. Specifically, these involve negotiations with agents. So certain memes find the human mind a more hospitable environment. The full story will have more complications, but the basic idea should be clear enough: religions are complexes of memes that are "designed" by memetic natural selection to flourish in the environment that was created in the human mind by evolution.

If any story of this kind can be made convincing, then any argument to the effect that religion itself cannot be explained naturalistically fails. However, a religious believer might still say that such a story could be true even if there is a God. Logically, that is certainly the case. But the believer can say a bit more, along the lines discussed above. Swinburne argues that a world governed by universal natural laws is what we would expect a benevolent God to create. But in such a world evolution by natural selection is the only way that living things, including us, can be produced. Similarly, it might be argued that the only way God can reveal the truth to people in such a world is through the processes described above: memetic selection in less than fully rational minds. This would not have to mean that God could not reveal the truth in any other way *at all*, but only that any other way would violate the laws of nature that God wisely ordained.

But a God that is so constrained, even voluntarily, is a very non-interventionist God indeed. Such a God is a far cry from the Judaeo-Christian one. It seems to be built in to the very idea of the Judaeo-Christian God that that God intervenes in the world by means of miracles, direct revelation to certain people and so on. The same can be said for the supernatural agents of other religions. A God who never did such things would seem to have far less to do than the gods of any traditional religions. Evolution does not provide a knock-down argument against religion, but it does seem to require a considerable watering down of religious beliefs as they have been understood through most of history. There may come a point where, if the watering down required is great enough, it will make sense to abandon all talk of God or the supernatural forever. The theory of evolution may yet kill God, but if it does it will most probably do so slowly.

13. Evolution and human nature

Much of the current public interest in evolution stems from the implications it has, or is alleged to have, for human nature. Darwin was aware of the possibility that it might have such implications: "In the distant future I see open fields for far more important researches. Psychology will be based on a new foundation, that of the necessary acquirement of each mental power and capacity by gradation. Light will be thrown on the origin of man and his history" ([1859] 1968: 458).

He took some steps in this direction himself with *The Expression of the Emotions in Man and Animals* (1872). Moreover, Freud attempted to give his work on the origins of religion and morals a Darwinian underpinning. Implausible though Freud's theories in this area seem today, one recent commentator has called Freud the "first evolutionary psychologist" (LeCroy 2000). In the mid-twentieth century, the idea that human beings have species-typical, evolved, "innate" mental characteristics fell out of favour in many quarters. This fall from favour was due to the influence of a number of different schools of thought, including Marxism, behaviourist psychology and the Margaret Mead school of anthropology. However, the evolutionary approach to psychology never completely disappeared. One of its manifestations was Desmond Morris's bestselling book *The Naked Ape* (1967).[1] Along with Morris's many subsequent books, it must take much of the credit (or blame) for fuelling the enormous public interest that has since developed in evolutionary approaches to human nature. This interest shows no signs of abating.

13.1 Sociobiology and its controversies

In 1975, Edward O. Wilson published a very large volume called *Sociobiology*, on the subject of group behaviour in animals. Its final chapter was about human beings, and became the subject of bitter controversy. The term "sociobiology" has since become synonymous with a particular approach to the study of human nature. This approach takes as its starting-point the fact that human beings are animals, and products of evolution. A core doctrine of sociobiology is that there are human psychological traits that are universal. This does not mean that they are necessarily possessed by every single person; rather, it means that they transcend cultural differences. Such traits, it is claimed, have remained largely unchanged since the Stone Age, and are adaptations to conditions at that time. Wilson believes that we can discover such traits by using adaptationist reasoning as a guide to research. That is, we can ask ourselves: what traits would it have been useful for Stone Age human beings to have? With our answers, we could frame hypotheses that could then be tested to see whether human beings actually do possess them.

Wilson's proposal can be seen as an application of the general principle that adaptationist reasoning can facilitate biological discoveries, a principle that I endorsed at the end of Chapter 4. Without the guidance of our evolutionary question, we would have no reason to research one possible trait rather than another. It would, of course, be bad scientific practice to infer that we actually possess a trait from the mere fact that it *would* have been useful in the Stone Age. By and large, neither sociobiologists nor their successors, evolutionary psychologists, have been guilty of this. Sometimes members of both schools have been speculative in their claims, but speculation is surely a necessary part of science, and only reprehensible if it is not clearly presented as speculation.

The adaptive traits that sociobiologists hypothesize are present in human beings are adaptive in the sense of promoting the replication of genes. Thus, sociobiologists can be seen to be endorsing the gene-centrist view of evolution that I discussed in Chapter 2. This leads them to claim that we should expect a higher level of altruism towards close kin than towards others. Some evidence has been produced in support of this. However, it might be thought to be an unsurprising finding, as we already knew that "blood is thicker than water". What sociobiologists want to prove, however, is that the relative thickness of blood and water is universal across cultures. If they succeeded in producing more robust evidence for this, they would have established a claim that is at least sufficiently interesting to have been denied by many anthropologists in the past. Sociobiologists have also investigated altruism towards non-kin. Indeed, it was Wilson who coined the useful terms "hardcore" and "softcore" altruism.

A further set of predictions stemming from the gene's-eye view of evolution concerns differences in behaviour between men and women. The key difference that generates the predictions is that a man can, in principle, father thousands of children, whereas a woman cannot give birth to anywhere near as many. Moreover, a woman is compelled to make a much larger investment in producing a child, having to incur opportunity costs for nine months as an absolute minimum. On the basis of this, it has been hypothesized that men and women have different priorities. From the point of view of genetic fitness, a woman should be choosy about who to mate with, whereas a man should seek to mate with as many women as possible. Further claims have been made about what should attract a woman to a man and vice versa. For example, it is predicted that women will value high status in men, whereas men will value youth in women. Hence they will be attracted to people bearing the visible outward signs of these things, respectively. Again, some evidence has been cited to support these claims, but they remain highly controversial.

When sociobiology first appeared on the scene, it attracted adverse criticism of a quantity and vehemence far greater than can be explained by the mere fact that it was a new and tentative theory. Wilson was subjected to personal attacks. On one occasion in 1978, he was about to read a paper at a symposium in Washington, DC. A group of protestors poured iced water over his head and, in classic student-demonstration fashion, chanted, "Racist Wilson you can't hide, we charge you with genocide!"[2]

On 13 November 1975, a letter appeared in the *New York Review of Books*, entitled "Against 'Sociobiology'". It was signed by sixteen people, including Gould, Lewontin and Ruth Hubbard. Here is a sample:

> The reason for the survival of these recurrently determinist theories is that they tend to provide a genetic justification of the status quo and of existing privileges for certain groups according to class, race, or sex. Historically, powerful countries or ruling groups within them have drawn support for the maintenance or extension of their power from these products of the scientific community ... Such theories provided an important basis for the enactment of sterilization laws and restrictive immigration laws by the United States between 1910 and 1930 and also for the eugenics policies which led to the establishment of gas chambers in Nazi Germany. (Allen *et al.* 1975)

On the face of it, the comparison between sociobiology and earlier "racial science" is hugely unfair. Sociobiologists (and evolutionary psychologists) have insisted from the outset that there is a *universal* human nature, in the sense of "universal" explained above. In support of this claim, they cite the well documented fact that genetic variation within the so-called "races" is far greater than between the most typical members of one race and another. In fact, perhaps ironically, Wilson cites Lewontin's own earlier work (e.g. Lewontin 1972) in support of this claim. It is open to question whether this genetic evidence actually supports the "universal human nature" claim. It seems to rest on the assumption that genes play a uniquely important role in determining pheno-type, an assumption that has been reasonably challenged by developmental systems theorists. Moreover, the idea of a universal human nature may itself be objected to on political grounds. However, given that sociobiologists make this universalist claim, they can hardly be accused of holding that one race is biologically superior to another.

The complaint that sociobiology has politically reactionary implications has refused to go away. One reason for this is the claim that human nature has not changed since the Stone Age and – barring genetic manipulation – is unlikely to change for a very long time. This suggests that certain persistent features of human life – such as war or the existence of social hierarchies – are things we are stuck with, however much we might dislike them. One might be tempted to derive the (mildly) optimistic conclusion that human nature is not going to get any *worse* either. This might mean, for example, that the sharp decline of religion in many countries in the past century is not going to lead to a "war of all against all". However, even this mild optimism is difficult to sustain in light of the fact that our weapons technology has "improved" vastly in recent times. If the tendency to make war remains what it has been, then we are liable to wreak greater destruction than ever before.

Many people, and not just those on the extreme left, like to believe that we can, through political reform or education, make the world better. Moreover, although it is not often said, political conservatives might make a case of their own for the potential mischievousness of sociobiology. A conservative might be troubled by a quietism that says: things will stay more or less the way they are no matter what we do. Presumably conservatives believe that established institutions play a necessary role in upholding desirable human traits, such as (perhaps) enterprise or fellow-feeling. They might not be prepared to trust to an evolved "human nature" to preserve these traits. Progressives, on the other hand, can point to the abolition of slavery and the rise of universal suffrage in support of their belief that we can improve things if we try. But a despairing view of human nature might lead us to dismiss these changes as superficial: in the West we have abolished slavery, but we still have wage-slaves; everybody has a vote, but power remains in the hands of small elites. Thus the suspicion that our own nature condemns us to never achieving genuine, non-superfi-cial, political change remains alive. Sociobiology was perceived by many, right from the beginning, as lending support to this despairing view. John Gray's book *Straw Dogs* (2003) takes the despairing view to an extreme, and claims to derive it from a Darwinian understanding of human nature: "though human knowledge will very likely continue to grow and with it human power, the human animal will stay the same: a highly inventive species that is also one of the most predatory and destructive" (Gray 2003: 4). But although he seems to be influenced by sociobiology, Gray is not himself a sociobiologist, and in fact takes Wilson to task for being unduly *optimistic*. One might defend Gray, and anybody else who uses sociobiology-like arguments in this way, by saying "Don't shoot the messenger". If the truth is that we are condemned, then we might as well know it. However, sociobiology's politically motivated critics argue that this despairing view is in danger of being a self-fulfilling prophecy: if we cannot change things, there is no point in trying to change them. Note, however, that this accusation only sticks if (a) sociobiology makes claims that actually do tend to support the despairing view, and (b) those claims are false, or at least ill-founded. Much criticism of sociobiology has been aimed at establishing (b). I shall address this question a little later.

It is worth mentioning a further response that sociobiologists and evolutionary psychologists have often made to the accusation that their work has political implications. This is to invoke Hume's doctrine that one cannot derive an "ought" from an "is". This response misses the critics' point, however, as it does not address the concern that saying that something cannot be changed is tantamount to saying that we should not bother trying to change it. One can reply to the invocation of Hume's doctrine with Kant's doctrine that "ought" implies "can". If we cannot make ourselves better, then there is no moral requirement that we do so. Moreover, as D. S. Wilson *et al.* (2003) have recently argued, saying that one cannot derive an "ought" from an "is" is *not* the same as saying that facts are *irrelevant* to deciding "ought" questions. I shall discuss these issues further in the next chapter.

13.2 Evolutionary psychology's grand synthesis

These "sociobiology wars" are far from over, but it is not because of them that a newer school of Darwin-inspired study of human nature has emerged. Although sociobiology is by no means dead, *evolutionary psychology* can be considered its successor theory.

13.2.1 The modularity thesis

Evolutionary psychology differs from sociobiology in explicitly endorsing the *modularity thesis* about the human mind. The modularity thesis has for a long time been a central pillar of cognitive science. It is the thesis that the mind is made up of a great number of special-purpose cognitive mechanisms, like the many blades of a Swiss Army knife. It may be contrasted with a view of the mind as composed of a few general-purpose mechanisms – for example, reasoning, learning by imitation – into which more or less any kind of content can be poured. On this latter view, reasoning (say) is completely neutral as regards topic: we can reason as easily on one subject as on any other.

On the modularity thesis, it is held that different parts of the mind operate in relative isolation from each other. That is, they do not always share information with each other. Each has its own specific domain wherein it operates. For example, those parts of the mind that process colours are dedicated specifically to that task. Modules are also characterized by *proprietary algorithms*, that is, the mechanisms they each use to go about their respective tasks may differ from each other. Moreover, the operation of modules is automatic. We do not choose to switch them on or not. Each module does its own thing in its own way, and does so automatically. It is often emphasized that the divisions between cognitive modules are *functional* divisions only, that is, they need not correspond to divisions between different locations in the brain. (Hence, I use the expression "parts of the mind" rather than "parts of the brain", although they are presumably in the brain.) On the other hand, if it turned out that they did so correspond, that would not damage the modularity thesis in the slightest. Whether the brain has physically discrete subsections does not matter to the modularity thesis. So empirical testing of any claim to the effect that there is a module for such-and-such a task, or of the modularity claim in general, should not consist in tests to discover whether different functions are performed by different areas in the brain. Finding that the same area of the brain was active in carrying out two different kinds of task would not disprove the claim that those two tasks were performed by different cognitive modules. The crucial test, rather, is whether these tasks are *functionally dissociable*, that is, whether one function can be impaired while the other is not. One of the most widely accepted examples of modularity involves our ability to recognize faces. In a condition called prosopagnosia, a person is unable to distinguish faces, even though, at least in many cases, the person's vision is otherwise unimpaired. This suggests that the ability to recognize faces involves cognitive subsystems that are not involved in seeing other things. We have, we might put it, mental software dedicated to recognizing faces.

A second source of evidence that a function is modular is where, even in normal people, we find discrepancies in abilities: specifically, where one kind of task is regularly performed more easily than another kind of task that is more complicated. Again, recognizing faces provides an example. Distinguishing faces is not, at least in any obvious way, any less complicated a task than distinguishing irregularly shaped rocks. Yet we find the former task the most natural thing in the world, and can recognize people we have seen only once before. It has proved difficult for computer engineers to get a machine to perform this task. Software used in closed-circuit television cameras for recognizing faces has been steadily improving, but still comes nowhere near matching human beings' effortless ability. The evolutionary reason for our having dedicated face-recognizing software is easy enough to guess. As social animals, it was presumably as important to our Stone Age ancestors to be able to tell each other apart as it is for us today. (Especially if those ancestors were practising reciprocal altruism.)

Evolutionary psychologists hold that our cognitive mechanisms were designed by natural selection to meet specific kinds of situation. They predict that we will find it much easier to think about these kinds of situations than about others for which our minds were not designed. A more controversial case than face recognition concerns the ability to reason using *modus tollens*. This is the principle that if we are given the premises "if *A* then *B*" and "not *B*", we can conclude "not *A*". It is well established that people perform poorly in the Wason selection task, which requires *modus tollens* reasoning. They still do poorly even if familiar, everyday examples are used. Cosmides and Tooby (1992) hypothesized that people had adaptive mechanisms for detecting cheating in others, rather than all-purpose logical reasoning abilities. This led to the prediction that people would perform significantly better in the Wason selection task when it was presented in a form that involved detecting cheating. That is, we find it much easier to draw the correct inference when the premises are of the kind "you must have a ticket to get in" than when they are of the kind "if Socrates is a man, Socrates is mortal". This prediction was subsequently confirmed experimentally. Cosmides and Tooby also claim that this generalization holds true whether the examples used are familiar scenarios or not. Specialized cheater-detection modules, they claim, would have been useful to our Stone Age ancestors, and their existence would explain the apparent discrepancy in people's reasoning ability. However, the interpretation of Cosmides's results has been disputed. Further experiments carried out by Dan Sperber and his team (1995) suggest that whether the cheater-detection scenario is a plausible one affects people's ability to perform the task.[3]

What evolutionary psychologists claim is not just that some functions of the mind are modular, but that virtually all of them are. This is known as the "massive modularity" thesis. A number of arguments can be advanced in support of this. Some are empirical, some are conceptual, and some are a mixture.

(i) Some evidence can be adduced for relative isolation of different functions of the mind from each other. Think of optical illusions. The lines in the well-known Müller–Lyer illusion appear to be different lengths. Even when one has measured the lines and seen that they are the same, the optical illusion does not go away. This apparently shows that the part of the mind that processes visual input does not receive all the information that is available to other parts of the mind. The knowledge that the lines are the same length does not seem to get through to the visual-processing mechanism; it still "thinks" they are different lengths. Evolutionary psychologists would add to this story that it is because evolution has not prepared us for this trick that the lines in the Müller–Lyer illusion appear different lengths in the first place. There presumably were no such lines around in the Stone Age.

(ii) A convincing case can be made based on general features of the phenome- nology of perception. Perceptual features of the world – colours are a good example – seem to just strike us. We just see that something is red. But the mind somehow works out that it is red, based on information received from the retina. As Pinker (1997: 7) points out, there is less light coming off a snowball indoors than off a lump of coal outdoors, yet to us the former looks white and the latter looks black. Somehow the information our retinas send to our brains is corrected to factor in background condi- tions such as lighting. But this goes on without our conscious knowledge. Some philosophers of empiricist leanings thought that, when we look at a round table, what we actually *perceive* is an oval shape, and that we *infer* the presence of a round object. In some sense this must be true, as an oval shape is what strikes our retinas. But it does not describe our experience truthfully. We are not consciously aware of making any such inferences. As anybody who has studied drawing knows, it takes special training to see the two-dimensional shapes that objects present to our eyes, and even then we can only do it for short periods of time.

(iii) Support for the modularity thesis also appears to come from the Chomskyan theory of how we acquire language. According to Chomsky, the actual input received by our ears and eyes contains too little information to enable us to acquire language using a general-purpose learning mechanism. A general tendency to imitate others would not be sufficient to explain how we acquire language. This is known as the *poverty of stimulus* argument. Chomsky concluded that we have inbuilt knowledge of how languages are structured. He further claims that all languages have the same "deep structure", so the inbuilt knowledge is universal.

Some linguists have disputed Chomsky's claims. For example, Geoffrey Sampson (1997) has argued that we can explain all the facts by positing an all-purpose trial-and-error learning mechanism. (He has been inspired by Popper in developing this view.) I do not propose to adjudicate this particular debate here. However, while Chomsky may not have right all the details of how language is acquired, the idea that all-purpose learning devices do seems to be a non-starter. This is because there is a general argument that is independent of the empirical details.

(iv) To see why I say this, try to imagine how a general-purpose mechanism would actually work. Suppose I pick up a glass, and then ask you to "do what I have just done". What should you do? Pick up a glass that is in front of you? Take hold of the glass I have just picked up? Or simply make the same motion of the hand that I made? The answer might seem obvious, but, whichever you choose to do, it goes beyond what is given in the instruction to "do what I have just done". We must interpret a person's behaviour as an *action of a certain kind*, if we are to imitate it. This is known as the "frame problem". In computer programming, it is the problem of reducing a set of potentially infinite choices to a finite – and hopefully reasonably small – number. Most of the time, the required interpreting goes on completely unconsciously. That is why, most of the time, following an instruction to "do what I have just done" does not present anyone with actual perplexities. (Incidentally, the frame problem presents yet a further problem for meme theory, over and above those I mentioned in Chapter 3.)

In order to acquire language at all, a person needs to be able to distinguish behaviours that are linguistic from those that are not. And in order to understand language, a person needs to be able to extract the *relevant* information from linguistic behaviour. Think of the word "glass" as pronounced by a Yorkshireman and a Londoner, respectively. They sound quite different, yet we are able to recognize that they are both tokens of the same word, whereas the Yorkshireman saying "brass" is not. And when the Yorkshireman asks me to repeat "the glass is made of brass", I know that he is not asking me to put on a Yorkshire accent. Once again, to imitate somebody, I have to interpret the behaviour as a certain type of – linguistic – action. True, we sometimes have difficulty understanding unfamiliar accents, but the task of adjustment is usually achieved. This task must involve highly sophisticated information processing. There is commercial software that enables me to train a computer to "recognize" words that I utter. But this is *dedicated* software. What all this suggests is that we too have dedicated "software" for processing linguistic information.

(v) A set of further arguments for the modularity thesis can be grouped under the heading of *evolvability* arguments. These are forcefully presented in Tooby and Cosmides (1992). Essentially, most of these arguments are the same as the general arguments for modularity that I discussed in Chapter 5, but applied to psychology. Adaptations arise in response to *specific* problems in *specific* contexts; and there are advantages to having mechanisms that are decoupled from each other. To these the evolutionary psychologists add a distinctive argument of their own (vi).

(vi) Dedicated mental mechanisms are *better* at their tasks than more multipurpose ones. If a situation demands a response immediately, it is best that that response be automatic. A mechanism whose operation consists in automatically responding to a particular type of situation is, it is argued, more efficient than a mechanism that can turn to various different tasks. The thought here is that the second type of mechanism would take more time deciding what to do.

You will doubtless have noticed that the specifically evolutionary arguments assume that the best explanation for how the mind came to be the way it is is adaptationist. It seems to be being taken for granted that the mind is largely a set of adaptations. It is all very well, it might be argued, to say that it is better adaptively to have a modular mind than an all-purpose one. But perhaps there are constraints that prevent this outcome from being achieved. This argument has been articulated by Paul Sheldon Davies (1999), and Todd Grantham and Shaun Nichols (1999). They argue, following Gould and Lewontin, that for all we know many features of the mind might not be adaptations at all, but spandrels or by-products. Grantham and Nichols add a further objection to what they see as the panglossian strain in evolutionary psychology. Many of the adaptive problems that evolutionary psychologists claim we probably have evolved mechanisms to solve were, they argue, not really very pressing ones. It might be *useful* to be able to "capture animals, acquire grammar, understand and make tools, or help relatives" (Grantham & Nichols 1999: 58), but it would not be *fatal* if one were not. Thus, they suggest, perhaps selection pressure just was not intense enough to produce the fine-tuned mechanisms that evolutionary psychologists claim we have.

I have already expressed scepticism regarding Gould and Lewontin's claims. Evolutionary psychologists have responded to objections of this kind by arguing that it is highly unlikely that non-adaptive forces could produce highly organized mechanisms. Note that the frame problem argument for modularity, and the other non-evolutionary arguments (i)–(iii), are untouched by the anti-panglossian arguments, since they do not rely on an appeal to adaptationist reasoning. But the frame problem arguments suggest that we do in fact possess highly sophisticated information-processing capacities, and that we could not have obtained these from experience. Adaptation by natural selection seems to be the only thing that could produce such highly sophisticated mechanisms (barring deliberate design, but I take it that this can be ruled out). They seem, that is, to pass the "functional analysis" test for being adaptations. If I have been underestimating the strength of Gould and Lewontin's arguments, then perhaps the adaptationist arguments for the modularity thesis are not all that convincing. But neither are they needed. The non-evolutionary arguments can stand up on their own. Once we have accepted them, then it seems plausible to say that the different modules are likely to be adaptations. Thus, evolutionary psychologists' quest to discover the problems our minds possess adaptations to solve seems a reasonable one. So there is, in my opinion, a strong case in favour of the modularity thesis. And there is also a strong case that cognitive modules are adaptations.

Another concern remains, however. Grantham and Nichols argue that adaptation may be constrained in such a way that the actual solutions it produces to problems may be less than perfect. So, even granting the evidence that we possess a host of cognitive adaptations, they may not be as well designed as they might be. Some statements by evolutionary psychologists seem to ignore this consideration: "From among the thousands of ways in which men differ, selection over hundreds of thousands of years focused women's preferences, *laser-like*, on the most adaptively valuable characteristics" (Buss 1999: 103, emphasis added). Clearly, Buss is assuming here that adaptation is pretty much unconstrained. The consideration of constraint may also weaken the adaptive argument for decoupling of cognitive modules. Perhaps the decoupling that *would* have the advantage of avoiding undesirable side-effects is not possible. The claim that the mind consists of numerous cognitive modules is different from the claim "one module per adaptive problem". Perhaps the truth is messier than this, as Griffiths (2007) suggests. Many of our physical organs perform more than one function, even though we can differentiate one organ from another. In the cognitive realm, our ancestors presumably used their visual-processing mechanisms for hunting, avoiding predators and mate-selection. Perhaps, then, various features of our visual processing are compromises between the different requirements of these different tasks.

However, the consideration of constraint just means that evolutionary psychologists should not reason from the mere fact that something *would* be adaptively advantageous to our ancestors to the conclusion that we actually possess it. But to repeat what I said earlier, I do not think evolutionary psychologists are on the whole guilty of this; they do try to test their hypotheses. Adaptationist reasoning does not by itself prove that we possess some trait, but it can help to cut down the number of empirical research programmes that we might decide to follow.

13.2.2 Advantages of evolutionary psychology over sociobiology

The modularity thesis is what distinguishes evolutionary psychology from sociobiology. What this means is that evolutionary psychologists claim that our inheritance from our Stone Age ancestors is a set of cognitive modules. Sociobiologists are not entirely clear as to what they take our inheritance to be. Sometimes it sounds as though they are claiming it is patterns of behaviour. If this were the case, it would mean that behaviours that were adaptive in the Stone Age would still be around now, albeit that they might not now be adaptive.

But it is *prima facie* the case that we do not behave in the same way as our Stone Age ancestors. Only a few of us hunt, while most of us drive cars, talk on telephones and use money. Sometimes attempts are made to convince us that distinctively modern behaviours are somehow the same as Stone Age ones. The man going to work on the Stock Exchange is a "hunter" aiming to "make a killing"; going to the hairdresser is "grooming behaviour"; and so on.[4] But such re-descriptions are insensitive to the differences between behaviours. Moreover, they do not tell us anything about what psychological mechanisms are involved in, say, learning to drive. On the other hand, saying that distinctively modern behaviours involve the same cognitive modules as our Stone Age ancestors had does not commit us to re-describing modern behaviours in this way. While sociobiologists can say that behaviours that were adaptive in the Stone Age may fail to be adaptive now, they cannot have much to say about behaviours that did not exist in the Stone Age. Evolutionary psychologists, by contrast, can go one better, and hold that cognitive mechanisms that produced certain behaviours in the Stone Age may produce different behaviours now.

Moreover, it is difficult to tell a plausible story about how genes – or genes in conjunction with developmental programmes – *could* shape human behaviour itself. The outcome of a process of development is a physical structure of some kind. If it is of a very simple kind – for example, a membrane that contracts in the presence of light – then its behaviour will be very easy to predict: it will contract in the presence of light. Human beings undoubtedly have *some* simple behavioural responses of this kind, such as the well known "knee-jerk" response to a tap on the knee. But the behaviour of a much more complex mechanism can only be described, if it can be described at all, by a long list of *if–then*, *but* and *either–or* clauses. One might argue that this is only a difference of degree, albeit a big one. But, first, *all* differences in biology are "only" differences of degree (as I hope I made clear in Chapters 8 and 9). But we do not call a bacterium the "same kind of thing" as an elephant. Secondly, the mediation of behaviour by a more complex mechanism is more *indirect* than a light-sensitive membrane.

Take, for example, the liking that human beings have for sweet-tasting foods. This is standardly taken to be an adaptation, as a moderate amount of sugar is good for us, and our Stone Age ancestors did not have the abundance of sugary foods that we have today. To eat as much sugar as possible would be a reasonable strategy for someone to whom not much sugar was available. But evolution has not bequeathed to us the behaviour of eating as much sugar as possible. Rather, it has bequeathed to us mechanisms that make eating sugar pleasurable and desirable. From the neurophysiological point of view, our evolved mechanisms cause sugar to trigger the release of chemicals in the brain. From the psychological point of view, our evolved mechanisms cause us to represent or experience sugar as something pleasurable and desirable. But many people do not eat as much sugar as they can get, and not just because they do not like it due to abnormal neurophysiolgical or cognitive development. Quite the contrary, many people who *love* sugar do not pig out on it. Thus, our legacy from the Stone Age is not the behaviour of eating as much sugar as we can get. It is the experience of sweet tastes in a certain way: as pleasurable and desirable. As Symons puts it:

In modern industrial societies, where refined sugar is abundantly available, the human sweet tooth may be dysfunctional, but sugar still tastes sweet, and the goal of experiencing sweetness still motivates behavior. That's how we're made. We can decide to avoid refined sugar, but we can't decide to experience a sensation other than sweetness when sugar is on our tongues. (1992: 139)

That sensation may be *explained* adaptively by the fact that it reliably produced a certain behaviour in the Stone Age. But the sensation is not the same as the behaviour; nor does it automatically produce the behaviour. The experiences we have just *strike* us: the rose strikes us as red, sugar strikes us as sweet, and certain individuals strike us as more attractive than others. But we are not *compelled* by these experiences to behave in certain ways; we do not inevitably eat all the sugar we can get, or try to mate with the attractive person. Evolutionary psychology's change of focus from behaviours to cognitive modules respects these facts.

But there is a commonly made accusation that evolutionary claims about human nature imply that we are controlled by our genes: that we have no choice about what we do. Gould, for example, articulates this accusation: "If we are programmed to be what we are, then these traits are ineluctable. We may, at best, channel them, but we cannot change them either by will, education, will, or culture" (1978: 238). The phrase "programmed to be what we are" is decidedly ambiguous. If it means "programmed to do what we do", then it is a caricature of what evolutionary psychologists are claiming. It would be closer to their view to say "programmed to experience the world the way we do". But surely our experience of the world *is* something we have no control over. Sugar tastes sweet, whether we want it to or not, and we cannot change that by will, education or culture. These experiences are, in Gould's usage of the word, ineluctable. That does not imply that our *behaviour* is programmed.

Admittedlly, rather than saying that our experiences are ineluctable *tout court*, we should say that the *basic constituents* of our experiences are. We have experiences of things such as cars, playing cards and love of football, and they did not exist in the Stone Age. Enculturation is undoubtedly needed to see a car as a car, or to experience love of football. But we still cannot help seeing a red car as red, solid and so on. The culturally specific items of our experience are made up out of basic constituents, which are the legacy of evolution. It is one of the central pillars of cognitive science, and hence of evolutionary psychology, that these evolved basic constituents are much richer in content and diversity than mere basic colours and shapes. Colours and shapes are *among* them, but so too are such things as "noun", "verb", "face", "sexual attractiveness" and perhaps categories such as "spider" or "dog". Doubtless, too, we can modify the way we experience things to some extent by means of self-discipline or therapy. We can, that is, learn to feel calm in a situation that had previously made us feel panicky. But the feelings themselves, calm or panicky, as well as happy, sad, angry, afraid, infatuated and so on are a set that we just have as part of our human nature.

I have perhaps been a little unfair to sociobiologists in criticizing them for taking behaviour to be what we have inherited from the Stone Age. The problem is that it is not clear what our evolutionary heritage is supposed to be according to sociobiologists. And sometimes it does sound as though they are saying that it is behaviour. For example, E. O. Wilson has said that: "the evidence is strong that a substantial fraction of human behavioural variation is based on genetic differences among individuals" (1978: 43). It is understandable that statements like this have been taken to be expressions of genetic determinism. The real problem, though, is that the phrase "is based on" is rather unclear. It is not clear that it is *not* genetic determinism. Evolutionary psychologists, by contrast, have made their position on this question clear.

A further difference between evolutionary psychologists and sociobiologists is that evolutionary psychology aims at a much broader range of targets. Sociobiology, remember, began life as the study of animal behaviour in groups. Specifically human sociobiology retains vestiges of its origins, in that the focus of interest of its practitioners has remained the evolutionary explanation of what happens when human beings get together. A vast amount of sociobiological literature is on the issues of altruism, mate choice and the spread of cultural trends, all of which have to do with how we interact with each other. Evolutionary psychology casts its net much more widely, as well it might, for the argument that if something is functionally complex then it is probably an adaptation surely applies to many features of the mind other than those to do with interpersonal relations. Our Stone Age ancestors surely faced a great number of adaptive problems other than social ones, for example, how to tell ripe from unripe fruit, or how to extract information about three-dimensional objects from two-dimensional arrays on their retinas. Evolutionary psychology aims at nothing less than a complete picture of the mind. Evolutionary psychologists have worked on colour perception (Sheppard 1992), and our responses to landscapes (Kaplan 1992; Orians & Heervwagen 1992), among other things. Pinker (1997: 528–38) even dares to speculate on the evolutionary origins of our responses to music. (To be fair, Pinker makes clear that he is being speculative, and that his explanation is missing some vital ingredient.)

An interesting consequence of evolutionary psychologists' overall picture of the mind is their stance on the "nature–nurture" question. As I indicated in Chapter 5, it is standard practice to say that the distinction between innate and acquired is meaningless, and evolutionary psychologists conform to this practice. But they have a further gloss on this, which is to say that calling something "learned" is a non-explanation. To see why they say this, recall my earlier point about why we cannot simply ask someone to imitate some action ("do what I have just done") unless that person is able to categorize the action: is able, that is, to tell what "doing what I have just done" consists in. Likewise, for anyone to learn anything at all, there must be already in their mind a way of interpreting the input from outside. The mind cannot simply passively absorb inputs. The way it processes any inputs it receives must be conditioned by already existing cognitive modules. Cosmides and Tooby express this point thus: "What effect the environment will have on an organism depends critically on the details of its evolved cognitive architecture. For this reason, coherent 'environmentalist' theories of human behavior all make 'nativist' claims about the exact form of our evolved psychological mechanisms" (1997). One might want to say that while this may well be true of the earlier stages of learning – to learn colour words, you need to already discriminate colours – that does not imply that it is also true of later learning. Surely, learning later on can involve the processing of information by means of concepts provided by learning earlier on? Thus, the objection might go on, the operation of evolved cognitive mechanisms need not condition learning at later stages.

But the evolutionary psychologists' point is more subtle than this objection allows. It is that the outcome of any processing of any input whatsoever is a joint product of the input and the cognitive mechanisms that process it. This means that if later learning is built on the outcomes of earlier learning, it is still conditioned by the cognitive apparatus that conditioned that earlier learning. Our evolved cognitive apparatus does not just help us to start absorbing information, and then go away. Every time some piece of information is used, it is used in the form in which our mind interprets it. This means that at no point does learning go on that is unmediated by our evolved cognitive apparatus. Note, incidentally, that this argument goes both ways: an array of cognitive mechanisms that received no input whatsoever would do nothing. (What goes on in people's minds in sensory deprivation tanks – the vivid hallucinations they reportedly experience – is no counter-example to this. The buzzing and humming of the brain's own physical activity is input.) So, just as we cannot analyse an organism into "innate" and "acquired" traits, nor can we analyse our minds' contents into "learned" and "unlearned". To say of some piece of knowledge, some personality trait or some behaviour that it is "learned" is to say nothing.

Although Cosmides and Tooby believe that scientific considerations support this conclusion, it can be reached without these, on purely conceptual, philosophical grounds. A mind that simply absorbs information without processing it is a conceptual impossibility. This is because any mind has to *represent* the things in the world in some way. Things do not simply wear the correct way to represent themselves on their faces. The mind must, therefore, bring something to the deal. It must already have some propensity to represent a thing one way rather than another. For Cosmides and Tooby (*ibid.*) to write that "coherent 'environmentalist' theories of human behavior all make 'nativist' claims about the exact form of our evolved psychological mechanisms" indicates that they are aware of this conceptual argument.

They are careful to avoid use of such terms as "innate" and "inbuilt" to describe the cognitive architecture that they believe we have inherited from our Stone Age ancestors. Their preferred terms are "evolved" and "species-typical". The cognitive mechanisms themselves are joint products of genes and environment. But so, too, are the anatomical traits that we all – or almost all – have in common: our hearts, livers and all the rest. (This is the starting-point of the developmental systems theorists' arguments, as we saw in Chapter 5.) We do not hesitate to say that hearts and livers are common to nearly all human beings, and that they are products of evolution. The high functionality of cognitive modules, Cosmides and Tooby consider, coupled with the long time it takes to produce adaptations, justify us in saying the same about them. They are not innate, because there is no such thing as innateness, but they are such as to reliably develop in a wide variety of different environmental circumstances:

> Because the world is full of potential disruptions, there is the perennial threat that the developmental process may be perturbed away from the narrow targets that define mechanistic workability, producing some different and nonfunctional outcome. Developmental adaptations are, therefore, intensely selected to evolve machinery that defends the developmental process against disruption (Waddington 1962). ... More generally, developmental programs are often designed to respond to environmentally or genetically induced disorder through feedback-driven compensation that redirects development back towards the successful construction of adaptations. (Tooby & Cosmides 1992: 80–81)

In other words, they are claiming that the evolved cognitive mechanisms are *canalized* (see Chapter 6). They even cite C. H. Waddington, one of the originators of the idea of canalization. Like the claim that the mind possesses such-and-such a mechanism because it is adaptive, the claim that such-and-such a mechanism is canalized must be put before the tribunal of empirical enquiry. Evolutionary psychologists are fully aware of this, and have attempted to justify their claims of universality by means of cross-cultural studies.

13.3 Conclusion

I have presented evolutionary psychology in a favourable light here. In fact, I have reservations about some of the claims made by evolutionary psychologists, and I believe that they have sometimes indulged in speculation without clearly indicating that they are so doing. For the most part, this has occurred in mass-market books that set out to popularize the field. But it is unfair to judge a scientific research programme on its popular manifestations.

Evolutionary psychology has attracted hostile criticism, much of it from the same people and in the same terms as sociobiology. However, I believe that much of this criticism is based on misunderstandings of what evolutionary psychology – in its non-popularizing manifestations – actually claims. For example, most of the criticisms in Rose and Rose (2000) are based on the assumption that evolutionary psychology claims that genes determine our behaviour. The criticisms of Sheldon Davies, Grantham and Nichols and others who make the accusation of "panglossianism" seem to me to be equally misguided, as I have argued. If I have seemed insufficiently critical of evolutionary psychology it is because I have been keen to rectify the misunderstandings on which many criticisms of it rest.

Although I have written that evolutionary psychologists have cast their nets more widely than sociobiologists, a very high proportion of their work – not just the popular presentations thereof – is about mating, parenting and group living, which are also the favourite subjects of sociobiologists. For example, in Buss's *Evolutionary Psychology* (1999), nine of thirteen chapters are on these subjects, and in a more recent large anthology of papers (Buss 2005), it is eighteen of thirty-four. It seems, then, that evolutionary psychologists have not entirely escaped the narrowness of focus of sociobiology. Sociobiologists, though, at least had the excuse that their discipline arose out of the study of animal behaviour in groups, and hence perhaps they should not be expected to aim at a comprehensive theory of the mind. Evolutionary psychologists, however, explicitly claim to have the wider aspiration. In their defence, I shall only say that the narrowness of focus is not an essential feature of evolutionary psychology. They could – and some do – turn their attention to less overtly sexy topics, while still pursuing a programme that was distinctively evolutionary psychological. Because of this possibility, evolutionary psychology is a promising and exciting research programme, which has the potential to yield a host of useful discoveries about the human mind.

14. Biology and ethics

In the immediate aftermath of Darwin's work, the idea that the theory of evolution had ethical consequences became quite widespread. Perhaps what makes this idea appealing is the popular notion that the theory of evolution in some way displaces or supplants religion. Religion is, after all, often seen as the source of moral teachings. If the theory of evolution significantly weakens the case for religious beliefs then, it might be thought, it had better generate moral teachings of its own. The alternative might be to be left with no moral teachings at all. However, there are a great many philosophers who think that ethics does not in any way depend on religion for its justification. Many of the early attempts to derive ethical consequences from evolution seem, from today's perspective, misguided. But the idea that we can learn something about ethics from evolution has not gone away.

14.1 Fitness as a normative concept

Evolution provides an account of how it is that living things have parts that are purposeful, that are *for* something. A world of purely physical entities, devoid of life, would presumably also be a world devoid of purposes. Those who are committed to a purely physicalist view of the world, then, are seemingly also committed to the view that purposes came into the world by means of evolution. Evolution explains why hearts, eyes and so on are for something. Moreover, it is the fact that they are for something that allows us to say that they are good or bad. A person's heart is good if it performs its function – pumping blood – well, and bad if it performs its function badly. Hence, those who are committed to a physicalist view can say that it is in virtue of evolution that things are good or bad.

But the heart's function of pumping blood itself has the function of keeping a person alive. It is the heart's purpose to pump blood *because* pumping blood keeps the person alive. A heart that pumped blood well, but at the same time had some other effect that endangered the person's health (maybe by producing some poisonous by-product) would not be a good heart. And we can push this further: keeping someone alive is the distal purpose of the heart because that in turn promotes evolutionary fitness. As we saw in Chapters 3 and 5, we might account for evolutionary fitness in different ways; it might mean replication of genes, or it might mean reproduction of the organism, or it might mean one of those plus group survival, and so on. Whatever evolutionary fitness is, however, it is that, and not any of the more immediate purposes of the heart, that its activity is finally directed to. Any of its more immediate purposes are its purpose only in so far as they serve this final one. Perhaps, then, we can give an account of what purposes things have, including human behaviour, in terms of evolutionary fitness? Perhaps, that is, we can find a basis for judging things good or bad, in the contribution they make, or the harm they do, to evolutionary fitness.

14.1.1 Evolutionary fitness as a basis for morals

The idea that we can is defended by William Rottschaefer and David Martinsen: "What makes an action good in the most fundamental sense in a DME [Darwinian meta-ethics] is that it promotes human adaptations and, thereby, fitness leading to S/R [survival/reproduction]. That is its ultimate justification" ([1990] 1995: 399). The use of the word "normally" here might look like a watering-down. But it just signals the fact that fitness is defined in terms of *expected* outcomes. A creature is fitter than another if it is so equipped as to be more evolutionarily successful, that is, to leave more offspring, propagate more copies of its genes, or whatever it might be. One creature, A, might be better equipped to reproduce than another, B, in that it is more fertile, or more attractive to the opposite sex. Assuming for the sake of simplicity that they are otherwise the same, one would *expect A* to do better than B. However, there is always the possibility that some accident could befall A, so that it leaves no offspring at all. Even if this were the case, it would still make sense to say that A was fitter than B.

To be fair to Rottschaefer and Martinsen, they write elsewhere in the same paper that "the robust Darwinian includes *among the multiple moral goods* those things and states of affairs that normally promote human adaptations, fitness and survival and reproduction" (*ibid*.: 393, emphasis added), implying that fitness is not the sole moral good. Nonetheless we must ask: is it plausible to think of fitness as a moral good at all?

Let us say that "pursue evolutionary fitness" was treated as a moral maxim. What kind of behaviour would it prescribe? It would not prescribe simple selfishness, as it would at the very least require us to look after our offspring, and to be prepared to make the ultimate sacrifice for them in certain circumstances. But the key term here is "in certain circumstances". Suppose a creature is able to distinguish between offspring that themselves have good reproductive prospects and those that do not. Let us say, for example, that some offspring are weak and sickly and have poor prospects of surviving long enough to produce many offspring. Or let us say that some of them are much better looking than others – by whatever the relevant standard of better looking is – so that it has a better chance of attracting mates. (In the case of a peacock, it would mean having a more elaborate tail.) From the perspective of evolutionary fitness, it would make sense to put more resources into the stronger, or the better looking, offspring, than into the weaker, uglier one. In an extreme case, putting resources into the weaker one is simply a waste: better to let it starve. Similar reasoning would apply to one's mate. In many species, the two parents share the job of providing for their offspring. Clearly, the two parents have a shared interest in seeing to it that their offspring are provided for. It is, therefore, in both their interests to stay together to look after their offspring. Once again, however, this is only in certain circumstances. The possibility could arise of finding a better mate: one who is more fertile, better able to provide resources for offspring, liable to produce offspring who themselves are healthier, or something along those lines. In such circumstances, the evolutionarily fit thing to do might be to leave one's current mate, literally, holding the baby.

We all know that things of this kind happen, or have happened. Moreover, we know that these kinds of things have been regarded by some as morally neutral or even commendable. The ancient Spartans are said to have put weak, sickly babies on a mountainside to die. And everybody knows that a common reason for marriage breakdown is that one partner has found somebody else.

14.1.2 Conflicting ends

I am not going to argue that these behaviours are simply immoral, as that would just be begging the question. I do point out, however, that the reason the Spartans left weak babies on a mountainside was not to promote evolutionary fitness. This should hardly be surprising, given that they had never heard of evolution. The reason the Spartans did this was, apparently, that they thought that the well-being of the city took precedence over the well-being of any individual. In the case of people leaving their partners for someone else, presumably only a very strange person does this out of a desire to promote evolutionary fitness. Evolutionary psychologists claim that the qualities we find attractive in someone, and the qualities that make us fall in love with someone, have adaptive explanations. That is, they are there because, at least for our distant ancestors, they were traits that made it fitness-enhancing to have that person as a partner. However, even if that is true, finding someone attractive is not the same as having the thought "it would enhance my fitness to have that person as a partner". The same point applies to one's attitude to one's own children. Once again, even if the phenomenon of parental love is explained by its contribution to one's own evolutionary fitness, that does not mean that one's attitude towards one's own children is "it enhances my fitness to take care of them". It is, surely, part of the very idea of *loving* a person that one has a concern for what is in that person's interests, that is, that one does not simply look after the other person's interests as a result of some calculation that doing so will serve one's own. Someone who was a pure moral egoist would not be able to take this stance towards anyone else. Similarly, someone who attempted to apply the maxim "pursue evolutionary fitness" would not be able to either.

All this shows, it might be argued, is that we *do not* actually follow this maxim. This, it might be argued, does not show that we ought not to. But the point has a little more force than this. The very idea of having ethics at all requires that some things be treated as having intrinsic value – or, to put it another way, that some things be treated as ends. The ends may vary from one person from one society to another. For example, most people today would think that an individual life has a value beyond its value to the state. And many people believe that marriage has a value that overrides the desires of one or other of the partners. Core ethical concepts such as good, right and duty all presuppose that there is something that is valuable in itself. Even utilitarians believe that maximizing happiness is a good in itself. At the very least, then, a whole host of ethical viewpoints presuppose that there is *something* that is a good in itself, that is, that deserves consideration independently of anything else. This does not mean that it cannot ever be sacrificed for anything else, for there may be several things that are ends in themselves that cannot all be served in every situation. One may have to choose between saving one person's life and saving another's, but that is compatible with each person's life being an end in itself. Or one may have to choose between telling the truth and preserving a life, even if telling the truth and preserving life are both ends in themselves.

The opposite of an end in itself is a mere means: something that is of value only because it is good for something else. Doubtless, we do think of many things in that way. A good overcoat is good because it keeps us warm in the winter. When we think of something as a means to some end, its value to us only goes as far as its serving of that end. Assuming that we value the coat *only* as a means to keep us warm, it is no longer valuable to us if it fails to do so, if it wears out, for example. Moreover, if some other coat becomes available that can serve the end better, we prefer it to the original one (once again, assuming that we value it *only* as a means to keep warm). Doubtless, too, there are many situations where we think of people only as means. If someone provides me a service – say, fixing my car – then I might no longer require that person's services when someone else comes along who can do a better job for the same price. I might, of course, have a sense of loyalty towards the person I have been dealing with for years, or he might be a friend, but in that case I am not viewing him as a mere means.

In the examples I gave above of people pursuing evolutionary fitness, they were treating other people as mere means. If you looked after your children only in so far as you believed that they stood a good chance of being reproductively successful, you would be treating those children as mere means. If you left your partner for someone else, because you believed that being with the new partner enhanced your evolutionary fitness, you would be treating both the old partner and the new one as mere means. One thing to notice about this is that it is not the way people normally think. But the further point is that, if one were to take "maximize evolutionary fitness" as a maxim, *everything*, apart from one's own evolutionary fitness, would be reduced to a mere means. Nothing, that is, would have any value except in so far as it promoted evolutionary fitness, and only as long as something else did not come along that promoted evolutionary fitness better. In some ways, this would be an attitude similar to that of the moral egoist, who also treats everything else as a means to the single end of his own well-being.

Some people might object to the very idea of moral egoism being called an *ethical* position at all. Kant argued that a moral maxim must be capable of being made into a universal law. One might object to egoism on the grounds that the maxim "always act so as to promote one's own self-interest" could not be made into a universal law. What would it be like if everyone acted that way? Everyone's moral maxim would conflict with everyone else's. Moreover, since everyone would only treat themselves as having intrinsic value, no two people would see the same thing as having intrinsic value. Doubtless it would be a very unpleasant world: not necessarily a war of all against all, but nothing more altruistic than a system of self-interested alliances. One would never know that one could trust anyone else. The same would be true of a world in which everyone was striving to maximize his or her own evolutionary fitness.

However, this objection is not decisive. When Kant argued that some maxims cannot be treated as universal law, he did not mean that the consequences would be *unpleasant* if we tried to make those maxims universal law. He meant that there would be something inconsistent in the action itself. For example, if I make a promise, I do so in the understanding that people expect promises to be kept. If there were no such expectation, I could not make a promise. But I can only break a promise if I have made one, so my breaking a promise depends on people in general having the expectation that promises will be kept. This is the sense in which promise-breaking cannot be treated as a universal law. Kant's point is not that life would be unpleasant if promise-breaking were generally accepted, but that it is *impossible* for promise-breaking to be generally accepted.

It is not clear that the same reasoning can be applied to the maxim "always act so as to promote one's own self-interest" or to the maxim "pursue evolutionary fitness".

14.1.3 Intuitions

The most obvious objection both to moral egoism and to treating "pursue evolutionary fitness" as a moral maxim is, quite simply, that it goes against so many of our moral intuitions. There are, undoubtedly, differences between cultures, and often what one culture regards as a moral duty another regards as an abomination. Most people today would regard putting babies out on mountainsides as an abomination. But that does not mean that there are no moral intuitions that are shared by all cultures, even our own culture and the Spartans. One candidate would be "murder is wrong". The Spartans presumably thought that putting babies on mountainsides was not murder. But it is possible for two people to disagree as to what, exactly, constitutes murder, and still agree that murder is wrong. The debate that still goes on – at least in some countries, such as the USA and Ireland – about abortion shows this. Those who oppose abortion think it is murder, while those who support it think it is not. Neither side is trying to persuade the other that murder is wrong or that murder is sometimes right. Rather, they agree that murder is wrong and disagree about whether abortion is murder. It is difficult to see how, without some weight being given to intuitions, moral arguments could get going at all.

I would suggest that the idea that one ought to treat some things other than one's own self-interest as having intrinsic value is an intuition of this kind. As evidence for this, I would point out that if we met someone who only ever considered his own self-interest, and viewed everyone and everything as a means to promoting this, we would regard him as seriously morally deficient. Although moral egoism is a possible moral position, in the sense of not being self-contradictory, our intuitions seem to tell us that it is *wrong* to view everyone and everything in the world as a mere means to one's own selfish ends, as wrong, perhaps, as murder. This would be so whether the selfish ends in question were promoting one's own well-being, or promoting one's own evolutionary fitness.

I wrote that intuitions carry *some* weight, but just how much weight do they carry? It may be that our intuitions clash with one another, so that at least some of them have to be revised or abandoned. But, whatever the theory of evolution tells us, it does not seem that the maxim that one should promote one's own evolutionary fitness could be an intuition at all. At the very least, this is because most people, at most times in history, had never heard of evolutionary fitness. It could no more be an intuition than "always stop when the light is red" could be. (That is not to say that one cannot *derive* the latter maxim from an intuition; perhaps there is an intuition "do not put other people's lives in danger unnecessarily".) But the claim that an intuition carries weight means that one ought to give some reason why it ought to be revised or abandoned. If something goes in the face of an intuition, as the maxim "pursue evolutionary fitness" seems to, then one ought to give some reason why it should be accepted. So it does seem that we cannot say that considerations of fitness by themselves can give rise to moral principles.

But one might ask: just what gives intuitions this weight? Even if they have some weight, can they *never* be overridden by anything else? Perhaps philosophical argument can override them. And perhaps scientific considerations can override them too. Moreover, what if there are situations in which our intuitions give us no guidance, or give us conflicting guidance? We might have no choice but to look elsewhere for answers. Again, perhaps science can help us. It seems to be saying an awful lot to say that scientific considerations should play *no role whatsoever* in moral deliberation.

14.2 The naturalistic fallacy

There are, however, quite a few people who say just that: scientific considerations should play *no role* in moral deliberation. At the other end of the spectrum from the claim that we can derive ethical norms from evolutionary fitness on its own is the claim that evolution, or indeed any scientific fact or theory, is *completely irrelevant* to settling any ethical question. Those who make this claim are not just people who believe that there are large areas of human existence that are forever beyond the reach of science. On the contrary, the claim is regularly made by evolutionary psychologists, one of whose defining characteristics is their ambition to produce a complete theory of human nature. For example, in their extremely controversial book *A Natural History of Rape*, Randy Thornhill and Craig Palmer write:

> A trait that increases this ability [to produce offspring that survive to produce offspring] is "good" in terms of natural selection even though one might consider it undesirable in ethical terms. There is no connection here between what is biologically or naturally selected and what is ethically right or wrong. To assume a connection is to commit what is called the naturalistic fallacy.   (2000: 6)

The reasoning can be summarized as follows. What the theory of evolution tells us is that living things acquired many, perhaps most, of the traits they have by natural selection. It is, then a *fact* about living things that they have traits that promote their fitness. But it does not follow that they *ought* to promote their fitness. To argue from a statement of fact to a statement that something ought to be the case is to commit what is known as the *naturalistic fallacy*. This expression was coined by G. E. Moore (1903), although the classic statement of the principle that one cannot derive an "ought" from an "is" was coined by Hume ([1739–40] 1978: bk III, pt I, §I) over 150 years earlier.

Hume asks us to consider inferences of the following type:

Stealing causes suffering to others.
Therefore, one ought not to steal.

Hume's point is that the above inference lacks a premise, namely:

One ought not to cause suffering to others.

What he is claiming is that a premise of this kind needs to be added to any "is" statement to derive an "ought" statement from it. That is, an argument with an "ought" statement as a conclusion needs to have at least one "ought'" statement as a premise. So we cannot derive an "ought" statement from "is" statements alone.

Moore's argument, and his conclusion, are a little different. What he is trying to show is that one cannot define "good" in terms of some other property. It is well known that he directs much of his wrath at utilitarianism. He argues that one cannot, as utilitarians do, define "good" as that which produces the greatest happiness. The argument he uses is known as the *open-question argument*. He argues that it, at least, *makes sense* to ask of a course of action that is known to produce the greatest happiness: but is it good? Contrast this with asking of someone who you have just been told is a bachelor: but is he unmarried? Assuming that you know and accept that the definition of "bachelor" is "unmarried man", it would make no sense to ask this question. Moore says that it *could* be the case that every course of action that produces the greatest happiness is good. But his point is that it is an *open question* – in the limited sense of a question that it makes sense to ask – whether an action that produces the greatest happiness is good. This shows, he concludes, that "good" cannot be *defined* as that which produces the greatest happiness.

With this weapon, Moore lays about him on all sides. Although he calls it the *naturalistic* fallacy, suggesting only that "good" cannot be defined in terms of *natural* properties, he also uses the argument against Kant. Kant's dictum that "good" can be defined as "that which one can will to be universal law" is called by Moore "the metaphysical naturalistic fallacy". Thus, what Moore is attacking is the idea that "good" can be defined *at all*. Rather, Moore thinks, we can only know what good is by intuition. Ultimately, any argument to show that something is good will be incomplete, as one of its premises will contain the term "good", which is indefinable. For example:

All actions that are X are good.
Giving to charity is X.
Therefore, giving to charity is good.

It could be that all actions that are X are good. But it could not be the case that all actions that are X are good *by definition*. To think that would be to commit the naturalistic fallacy. To find out whether all actions that are X are good, one would have to consult one's intuitions. Alternatively, one could take a shortcut and consult one's intuitions about the conclusion. There is no guarantee that either of these consultations will yield an answer. I might consult my intuitions about whether voting "yes" in a particular referendum is good, but my intuitions may not tell me either way. Moore himself is very cautious about what his intuitions tell him, finally concluding that the only things he feels fully confident in calling good are friendship and the apprehension of beauty.

It is very common for sociobiologists and evolutionary psychologists to invoke the naturalistic fallacy as a way of deflecting the criticism that their claims have unacceptable ethical consequences. All they are saying is that humans are such-and-such, and that being such-and-such is a consequence of natural selection, and hence is fitness-promoting, or at least was for our distant ancestors. Note that there are two types of claim here. First, it is clamed that there is a universal human nature, that is, that human beings in all cultures (although there may be individual exceptions) have such-and-such cognitive traits. Secondly, it is claimed that these traits are likely to have been fitness-promoting in our Stone Age ancestors. But these are both factual claims. To derive any "ought" statements from either of them, they argue, would be to commit the naturalistic fallacy.

How do Hume's and Moore's arguments apply to the specific issue of using evolution to justify value claims? Hume's argument can be applied in the following way. Suppose someone says "it is to our adaptive advantage to behave in such-and-such a way". To get from this to "we ought to behave in that way", we would need to supply the further premise "one ought to behave in ways that are to one's adaptive advantage". Even if this premise were true, it could not be supplied by science. Likewise, to get from the premise "most people like to behave in such-and-such a way" to the conclusion "one ought to behave in such-and-such a way", one would need to supply the premise "the way most people like to behave is the way one ought to behave". Once again, even if this is true, it is not something that can be discovered by science. In each case, then, the missing premise has to be supplied from somewhere else. This "somewhere else" may be intuition or philosophical reasoning, for example. There is no guarantee – indeed it seems extremely unlikely – that these sources will actually justify the premises "one ought to behave in ways that are to one's adaptive advantage" or "the way most people like to behave is the way one ought to behave".

Moore says that we cannot define "good". That means that we cannot define it as the way people in fact (mostly) are. The open question argument seems to work here. We can accept that most people actually have such-and-such a property, and it would still make sense for us to ask: but is it good that they have that property? Likewise, we cannot define "good" as that which promotes evolutionary fitness. We could accept that being such-and-such promotes evolutionary fitness, and it would still make sense to ask: but is being such-and-such good? If all we accepted was that being such-and-such promoted fitness for our distant ancestors, it would even more clearly make sense to ask: but is being such-and-such good? So far so good, it seems.

Moore conceded that it *could* turn out that that which produced the greatest happiness was always good. Likewise, for all that the open question argument tells us, it could turn out that the way most human beings are is good. The only way we could discover this, on Moore's account, is to consult our intuitions. I suspect that our intuitions would not give any answer until we had filled in some detail about how most human beings in fact are. We are, at least, familiar enough with the pessimistic thought that many people have bad traits, to be reluctant to give a blanket endorsement to the way most people are.

Let us say for the sake of argument that most people like to give to charity. This, our intuitions might tell us, is good. But when we consult our intuitions about this why do we have to go through the detour of finding out if most people actually do have this trait? Is it not just as likely that our intuitions will tell us that giving to charity is good, regardless of whether most people like doing it or not? Conversely, let us say for the sake of argument that most people would cheat on their partners if they thought they could get away with it. What do our intuitions tell us in this case? I suspect that they tell us that cheating on your partner is bad. If they do, it is not clear that discovering that that is how people in fact are, or even discovering that there is an adaptive rationale for people being that way, would give us any reason to abandon that intuition. In the case of giving to charity, if evolutionary psychology, or any other scientific theory, tells us that most people like to do this, it is not clear that that would give us any further reason to do it. It is not clear, that is, that it would supply any justification for doing it over and above what our intuitions already told us. Moore's point, it seems, still holds.

It could conceivably turn out that what is good coincides exactly with how people in general like to behave, or with how it would be adaptive for people to behave. But even if one does not accept Moore's further claim that only intuitions can tell us the answer, the open question argument still applies. We would still find out what was good by some means other than finding out how people in general like to behave, or how it would be adaptive for them to behave. Whatever the other means is – let us call it "ethical reflection" for want of a better term – if it is capable of answering questions about what is good at all, it is capable of doing so without knowing what is adaptive or what most people like to do. So, even if what is good coincides with, say, what is adaptive, we should not need to go through the detour of finding out what is adaptive to find out what is good.

But, as Wilson, Dietrich and Clark (2003) argue, there is a big difference between claiming that one cannot settle value questions using facts alone, and claiming that facts are *irrelevant* to settling value questions. In fact, neither Hume nor Moore make this more extreme claim. Hume draws the milder conclusion that empirical facts *by themselves* cannot tell us whether something is good. Consider the following argument:

> One ought not to cause suffering to others.
> Therefore, one ought not to steal.

This argument is just as clearly missing a premise as the one I gave above in my initial presentation of Hume's position. It is, moreover, an *empirical* question whether something causes suffering to others. Of course, in many cases we know perfectly well, without the aid of science, what causes suffering. We do not need science to tell us that hitting people, or putting them in prison, is likely to cause them suffering. But science may possibly tell us that things that we did not suspect caused suffering actually do. It is conceivable, for example, that science could discover a causal link between young children being treated in a certain way and emotional suffering in adult life. If we already had a premise such as "one ought not to cause suffering to others" then we would have a valid argument to the conclusion "one ought not to treat young children in that way".

It may be that the most general ethical truths – such as, if it is an ethical truth, that one ought not to cause suffering – are reached by non-empirical means. But when it comes to considering more specific courses of action, the foregoing argument suggests that empirical facts play a role after all.

Moore leads us to a similar conclusion. He claims that friendship and the apprehension of beauty are good. If they are, then a world with more of either of these things would be better than one with less, if the two worlds are otherwise the same. This means that any act that increases the total amount of one of these goods without decreasing the amount of any other good is a good act. Hence, although facts play no role, for Moore, in deciding whether a thing is good as an end in itself, they do play a role in deciding whether an act is good. Thus he distinguishes two questions: "What kind of things ought to exist for their own sakes?" and "What kind of actions ought we to perform?" It is only the first type of question that he thinks can be answered by intuition alone:

> As for the *second* question, it becomes equally plain, that any answer
> to it *is* capable of proof or disproof – that, indeed so many different

considerations are relevant to its truth or falsehood, as to make the attainment of certainty impossible. Nevertheless the *kind* of evidence, which is both necessary and alone relevant to such proof and disproof, is capable of exact definition. Such evidence must contain propositions of two kinds and of two kinds only: it must consist, in the first place, of truths with regard to the results of the action in question – of *causal* truths – but it must *also* contain ethical truths of our first or self-evident class. (Moore [1903] 1993: 34)

One of Moore's own examples of something that is clearly – intuitively known to be – good, is friendship. Given that that is the case, then presumably it is good to *promote* friendship. But to do that, we need to know *how* to promote friendship. Let us say you are trying to get two people who are enemies to become friends. A little psychological knowledge would not go amiss. Your efforts to promote friendship would probably be a complete waste of time if you did not know what actions of yours were likely to actually promote friendship. Some of this empirical knowledge is doubtless just common sense, but it is at least conceivable that the science of psychology could provide useful information here. Thus, in order to decide what exactly would be a good thing to do, it would not be enough to know that friendship is good: some empirical knowledge would be required as well.

Those who still want to defend the more extreme claim could dig their heels in at this point. They might insist that our intuitions do not just provide general principles, but tell us what is good to do in specific situations as well. But this is highly implausible. A large reason for this is that new moral problems arise all the time, and it is unlikely that we are provided with intuitions for dealing with all of them. Some people think that one should not use disposable nappies, on the ground that they damage the environment. Even if it is our intuitions that tell us we should not damage the environment, surely it is an empirical matter whether or not disposable nappies do so. It seems extremely unlikely that intuitions, or any other purely non-empirical sources of knowledge, could by themselves provide an answer to the question of whether or not we should use disposable nappies.

The example of whether something causes suffering and the example of how to promote friendship suggest that psychological knowledge often has a role to play in deciding whether a particular course of action is good. So we cannot rule out the possibility that future psychological discoveries will help us in deciding specific ethical questions. Evolutionary psychology offers hope of discovering previously unsuspected features of human psychology, so we cannot rule out the possibility that it could help us to decide some ethical questions. The extreme position seems to depend on the implausible claim that intuitions by themselves can guide us on every specific question of what we should do.

The more moderate position actually held by Hume and Moore only says that empirical evidence by itself cannot decide these questions. It does not say that empirical evidence is irrelevant.

14.3 Ought implies can

There is a further way in which empirical considerations might be relevant to ethical questions. This can be seen if we think again about the worry that some people have about the fatalistic implications that they believe sociobiology and evolutionary psychology to have. The worry is that these disciplines might lead us to be more forgiving towards certain undesirable behaviours, by making it seem that people are genetically determined to behave in that way. Much of the hostile reaction to sociobiology and evolutionary psychology that we looked at in Chapter 13 stems from a belief that it gives people an excuse for certain actions. Both sociobiology and evolutionary psychology claim that there is a universal human nature that is a legacy from our prehistoric ancestors and is unlikely to change any time soon. Moreover, they claim that war (Buss 1999: ch. 9), male chauvinist attitudes (Wilson & Daly 1992) and the tendency to form social hierarchies (Buss 1999: ch. 12) all arise from that universal human nature. The concern of many of the critics, then, is that this implies that we are stuck with these things. As regards applying this to ethics, two consequences might seem to follow.

First, it might seem to follow that there is no point in making any effort to eradicate them. Feminists, it might be argued, are wasting their time trying to persuade men to change their ways. Pacifists are wasting their time trying to persuade people to solve their disagreements without resorting to war.

Secondly, it might seem to follow that people are not to blame for certain traits or actions. Men are not to blame for trading in their wives for a newer model. Governments are not to blame for failing to promote social equality. Stephen Rose expressed the idea in his review of Wilson's *On Human Nature* (1978): "for Wilson human males have a genetic tendency towards polygyny, females towards constancy (don't blame your mates for sleeping around, ladies, it's not their fault they are genetically programmed)" (Rose 1978, quoted in Dawkins 1982: 11).

If it was really, literally, true that a man was unable to stop himself from philandering, then it really would be wrong to blame him for it. For that matter, if someone was literally unable to prevent himself from murdering people, then it would be wrong to blame him for it, although it would probably be a good idea to lock him up. This is known as the principle of "ought implies can". The principle is generally attributed to Kant, but one does not have to buy into Kant's entire ethical system to accept it. Consider the following example: it would, let us say, be good if we could bring John Lennon back to life. Unfortunately, we cannot do so, so it would seem wrong to say that we *ought* to bring him back to life. Another way of putting this is to say that we are not under any obligation to do what is impossible for us to do. So, although empirical findings about how human beings are cannot, by themselves, *justify* any "ought" claim, it is conceivable that empirical facts about how human beings are could by themselves *falsify* an "ought" claim.

By the same token, if it is literally impossible to stop people from going to war, then pacifists, and for that matter political leaders who decide to try negotiation instead, really are wasting their time. Since time is a limited resource, they would be better advised to do something that might actually have an effect. They might, indeed, even be said to be morally to blame for so wasting their time. If King Canute had spent a considerable time trying to stop the tide, he would have been using time that could have been spent doing something more useful for his subjects, which entails doing something that stood some chance of actually having an effect. Even if stopping the tide would have been desirable – because there was a threat of flooding, say – he would still have been wasting his time. His subjects would have been perfectly justified in wishing for a better king.

However, this principle can only be used if it actually is literally impossible for men to stop themselves from philandering. It is literally impossible for me to grow wings, or to bring John Lennon back to life, so I cannot be under an obligation to do either of these things. But if it is not literally impossible for me to do something, then I cannot use the "ought implies can" argument to claim that I am under no obligation to do it. If it is not literally impossible for men to avoid philandering, then philandering men cannot use the "ought implies can" argument to absolve themselves from blame. But, as I argued in Chapter 13, it is a caricature of evolutionary psychology to say that it implies that men cannot help being philanderers because of their genes. This is because evolutionary psychologists claim that what reliably develops in many different environments is not behaviours but cognitive mechanisms. This includes the fact that when we see something red it looks red, and when we see a face we recognize it as a face. It also includes pleasurable sensations and desires. We get pleasure from sweet foods and mild, sunny weather, and we desire sexually attractive members of the opposite sex. That does not mean that we cannot help throwing ourselves at sexually attractive members of the opposite sex, any more than it means that we cannot help gorging ourselves on chocolate gateaux. Our legacy from our evolutionary past is desires, not behaviours. Rose's remark was directed against Wilson, a sociobiologist, but the worst that can be said about sociobiologists in this regard is that they do not make it entirely clear whether they are genetic determinists or not.

But more needs to be said. Rose's criticism (and there are many more people who say exactly the same thing)[1] presents things in a very all-or-nothing way. If men cannot help philandering because of their genes, then they are not to blame for philandering. But, if men can help philandering, does that mean that sociobiological and evolutionary psychological findings are irrelevant to ethical questions?

What about situations where it is just *very hard* for me to do something, or to avoid doing something? We might say that there is a trade-off between how difficult something is and how great the obligation is on us to do it; that is, it is not that we are only obliged to do easy things, but that if we have an obligation, however strong, on us to do something, that obligation is – or can be – lessened by the fact that it is extremely hard to do. At worst, evolutionary psychology tells us that men will forever be tempted to philander. It does not mean that they are unable to prevent themselves from doing so. But the worry that this provides an excuse for these behaviours, and that it means certain attempts at social reform are futile, is not entirely eliminated by this. It is *easier* for me to avoid doing something I have no strong desire to do. Perhaps, too, it is easier for social reform to bring about some changes than others. In criminal trials, the defendant's emotional state is often taken into account. If the defendant was under extreme stress at the time of the crime, that can be seen as a mitigating factor. If someone has a mental illness that makes it hard – not necessarily impossible – to avoid doing certain things, that can be seen as a mitigating factor too. Mitigation comes in degrees. Depending on just how hard it is for the person to avoid the action, we might decide to give them a reduced sentence, or to give them no sentence at all. There is the further complication that if somebody *continuously* finds it *extremely* difficult to avoid doing something, they might need to be locked up to protect others. But that is consistent with the person not being held to be to blame for their actions.

But it is not clear that evolution tells us *just how hard* it is to do anything. Is it plausible to claim that it is extremely hard for men not to philander? Many men seem to manage to restrain themselves. True, there are people called "sex addicts" who find it extremely difficult to stop themselves. But the very fact that we call these people "addicts" implies that we recognize that this characteristic is not normal. Likewise, some societies do better at promoting equality than others. Most modern states have social welfare programmes, recognize the right of workers to strike, and penalize employers for sexual or racial discrimination. Human nature *might* mean that we can never achieve full equality (assuming it is clear what that means) but it does not mean that all efforts to increase equality are futile. Likewise, human nature *might* mean that we will always have wars, but it does not mean that negotiation has never prevented a war. This all seems to fit with what evolutionary psychology says.

Evolutionary psychology tells us that people have desires that we may wish they did not have. At most, this might mean that they are less to blame for certain actions, or that the absolute ideals of no war and no inequality what-soever are unattainable. Marxists and 1960s anthropologists may not like this, but it is not news. It is, in fact, essentially the same as the Christian teaching of original sin, minus the biblical story of how we got to have original sin, which the Catholic Church tells us we are not supposed to take literally anyway. Believing in original sin does not entail believing that Adam and Eve ate an apple; it entails believing that human beings are originally sinful. Once again, Chesterton seems to have hit the nail on the head:

> Men thought mankind wicked because they felt wicked themselves. ... The only thing we all know about that primary purity and innocence is that we have not got it. Nothing can be, in the strictest sense of the word, more comic than to set so shadowy a thing as the conjectures made by the vaguer anthropologists about primitive man against so solid a thing as the human sense of sin. By its nature the evidence of Eden is something that one cannot find. By its nature the evidence of sin is something that one cannot help finding.　　([1908] 2004: 79)

Christianity can hardly be accused of claiming that people have no control over their actions, or of being morally lenient, just because it has the doctrine of original sin. What the doctrine might imply, and probably does in Christianity, is both that we should be understanding of those who sin (not the same thing as absolving them from all blame), and that we should be extra vigilant about certain things. If people are innately disposed to go to war, maybe that implies that we should *increase* our efforts to prevent wars, not give them up.

But the fact that this doctrine has been around for thousands of years surely suggests that, if that is all evolutionary psychologists are telling us, then they are telling us something we already know. Moreover, even if it is not strictly deterministic, it still sounds pretty pessimistic. I shall return to these points in §14.5.

14.4 Altruism

Much attention has been given to the question of whether human beings have innate altruistic dispositions. Evolutionists have speculated on whether a plau-sible evolutionary scenario would allow such dispositions to evolve. The general consensus is that indiscriminate altruism is not likely to have evolved in human beings, but that kin altruism and reciprocal altruism are. Much of the appeal of group selection stems from the guarantee it seems to give that what we might call *genuine* altruism – that is, hard-core, non-kin altruism – can evolve. Although group selection has lost its plausibility in many people's eyes, people have not given up on the idea that genuine altruism could evolve. Neither reciprocal nor kin altruism can be called altruism as it is usually understood. If true altruism is defined by the maxim "do unto others as you would have them do unto you", then reciprocal altruism is "do unto others as you expect them to do unto you", and kin altruism is "do unto your relations as you would have them do unto you, and even then only in proportion to how closely related they are". Can we get genuine altruism out of these two strategies?

Much of the heat generated by the debate about group selection seems to be generated by a concern that, if group selection is not possible, then we human beings, as products of evolution, are condemned to never be true altruists. This statement is, as it stands, too strong, as it clearly relies on the strong determinist understanding of genetics that, as I pointed out in the previous chapter, is disavowed even by adherents of the Hamilton–Williams view. However, the idea that human beings are *innately* disposed to altruistic acts is apparently appealing to many, and hence it might be encouraging to think that this is not rendered unlikely by the fact of our being products of evolution. Recall the distinction between evolutionary altruism and psychological altruism that I mentioned in Chapter 3. Behaviours that increase the fitness of other individuals at the expense of the fitness of the individual carrying out the behaviour are instances of *evolutionary altruism*. When an individual does something with the intention of helping another, this is an instance of *psychological altruism*. Williams's arguments against group selection suggest that evolutionary altruism is very unlikely to evolve. The upshot seems to be that psychological altruism is equally unlikely to evolve, since, surely, the *outcome* of psychologically altruistic acts would usually be evolutionarily altruistic. So, it seems, we are unlikely to have innate altruistic dispositions. Perhaps there is a way out of this, however. A number of ways in which evolution might produce innate altruistic dispositions even without group selection have been proposed.

Proposal 1: error theory
The first is a development of the kin-altruism idea. Bear in mind that what is evolutionarily relevant is the *outcome* of one's behaviour, not the motivation (if there is any) behind it. What is evolutionarily unstable is *evolutionary* altruism, not necessarily psychological altruism. If indiscriminate psychological altruism reliably *had the effect* of promoting the fitness of others in either of the discriminating ways just given, it could be an evolutionarily stable strategy. Michael Ruse suggests this when he writes: "Literal, moral altruism [i.e. psychological altruism] is a major way in which advantageous biological cooperation is achieved. Humans are the kinds of animals which benefit biologically from cooperation within their groups, and literal, moral altruism is the way in which we achieve that end …" ([1986] 1995: 229). But how could this happen? One way would

be if the creatures one regularly encountered were in each case *highly likely* to be close kin. For example, a gregarious species might form herds that consisted of kin groups. This would mean that any random individual one encountered would be highly likely to be a relative. This in turn would mean that by helping that individual one would be likely to be promoting one's own genetic fitness. As regards human beings, we know that that is not the state of affairs now, but it is generally agreed that natural selection is a very slow process, so the equipment bequeathed to us by evolution is not necessarily fitness-enhancing in the modern world. What we should expect is that it would have been fitness-enhancing in the *environment of evolutionary adaptedness* for human beings, that is, the situation in which human beings lived for most of the species' existence. (Similarly, domestic cats stalk birds even though they are fed by their owners every day, because in their environment of evolutionary adaptedness, stalking birds was necessary for them to live.) Let us hypothesize, then, that in the Stone Age human beings lived in kin groups. There would be no need for them to evolve mechanisms that discriminated between kin and non-kin: indiscriminately altruistic behaviour would reliably benefit one's kin. Since we have inherited those early human beings' innate dispositions, we have retained the disposition to be indiscriminately altruistic. This theory is referred to as an "error theory", as from the point of view of genetic fitness, which would be the reason the altruistic disposition was there, the altruistic behaviour would be an error. However, this scenario appears to achieve the desired result: a possible evolutionary story that can produce an innate disposition to altruism.

Unfortunately this theory runs up against some serious empirical problems. Hamilton's rule tells us that even in a kin group it makes sense to discriminate nearer from further kin, siblings from cousins, for example. Perhaps some simpler organisms are unable to make such discriminations, or perhaps the cost of developing the mechanisms for making them is, in fitness terms, too great. However, it is well established that chimpanzees, our closest relatives (jointly with bonobos), are able to recognize each other's faces, and can discriminate their relations from non-relations and from each other (Parr & de Waal 1999; Vokey *et al.* 2004). So it is likely that our Stone Age ancestors were able to do this too. Moreover, there is no evidence that Stone Age human beings actually lived in kin groups. So all the proposal does is show us a *possible* scenario in which indiscriminate psychological altruism could evolve. But that scenario relies on specific empirical conditions being met. The evidence suggests that these conditions were not in fact met in our environment of evolutionary adaptedness.

Proposal 2: the game of life
A second proposal does not give us absolutely indiscriminating altruism, but slightly softens the discrimination of reciprocal altruism (Axelrod 1990). To see how it works, let us first imagine that you make a deal with somebody to leave a

bag of money in a secret location, in return for which the other person will leave a precious painting in another secret location at the same time. Let us imagine that you both have to collect at the same time, that neither of you knows who the other is, dealing entirely through intermediaries, and that you will never have any dealings with each other again. This is a version of the *prisoner's dilemma*, a famous thought experiment devised by Merril Flood and Melvin Dresher in 1950.[2] Assuming that you are motivated only by self-interest, what is the rational thing to do? You do not know what the other person will do, but if you leave the money and he does not leave the painting in the secret location (i.e. if you *cooperate* and he *defects*), there is nothing you can do about it, so you have lost your money for nothing. On the other hand, if you both defect, then you have lost nothing, even if you have gained nothing either. And if you leave no money and he leaves the painting, then you get the painting for free. So it seems that the most rational thing to do is defect, since no matter what the other person does, you are better off defecting. But herein lies a paradox. If your opponent thinks this through, he will defect as well. But if you both defect, you both gain nothing, even though you both would have got what you wanted had you both cooperated.

Part of the reason for this paradoxical outcome is that you will have no further dealings with each other: it is a *one-shot prisoner's dilemma*. Things change if you have to deal with the same person over and over again: an *iterated prisoner's dilemma*. I shall pass over this, however, and move on to a third scenario. Let us now imagine that there are many individuals playing prisoner's dilemma-type games with each other. You only play with (or against, if you prefer) one other person each time, but let us say it is a matter of chance who you find yourself playing with each time. One other factor needs to be added: you are able to recognize players you have met before, and remember what they did before. (Remember that one of the problems with the "error-theory" proposal was that it required that individuals be unable to recognize each other.) What Robert Axelrod did was run a computer simulation of many individuals doing this, with suitably weighted rewards and penalties. Each player starts with the same number of points, and stands to gain or lose points with each exchange. The reward for defecting while the other player cooperated was bigger than the reward for cooperating while the other player cooperated, which in turn was bigger than the reward for defecting while the other player defected. The worst thing of all to happen was to cooperate while the other defected: this carried a penalty. (Looking again at the one-shot prisoner's dilemma should make clear why the numbers are set up this way.) Axelrod invited people to send in various strategies for individual players in this game. Examples include "always cooperate", "always defect", "cooperate at first and start defecting after ten exchanges with the same individual", "cooperate and defect equally often, but at random", and many other more complicated strategies.

The most successful strategy turned out to be a very simple one, called "tit-for-tat". This consists of cooperating whenever you meet an individual you have not met before, and thereafter treating that individual as that individual treated you the last time. Axelrod reasoned that the reason for tit-for-tat's success was that it was *nice* (cooperates as a default), *retaliatory* (defects against a previous defector), *forgiving* (if a previous defector starts to cooperate, so does tit-for-tat), and *clear* (easy for other players to work out what strategy is being used – this last indicating that its simplicity is one of its advantages): "Its niceness prevents it from getting into unnecessary trouble. Its retaliation discourages the other side from persisting whenever defection is tried. Its forgiveness helps restore mutual co-operation. And its clarity makes it intelligible to the other player, thereby eliciting long-term co-operation" (Axelrod 1990: 54).

A player using the tit-for-tat strategy (a tit-for-tatter) can suffer the penalty for cooperating while the other player defects only once against any player it meets. By the same token, the defecting player can only get the maximum reward out of an exchange with a tit-for-tatter once. On the other hand, if a tit-for-tatter meets either a cooperator or fellow tit-for-tatter, each gains the benefits of mutual cooperation. Even if these are not as high as the benefits of defecting when the other player cooperates, they are still better than mutual defection.[3]

Axelrod next devised a variant on this game, called the "game of life", which was aimed at modelling natural selection, by having survival and reproduction be affected by the number of points one amassed. We can call creatures that always cooperate *altruists*, and those that always defect *egoists*. If the population consisted entirely of altruists, each of their points would gradually increase, and so their numbers would gradually increase too. But we can introduce some egoists to the picture. Given that the altruists are indiscriminate, they will cooperate with the egoists just as much as with each other. The upshot will be that the egoists reproduce faster than the altruists, and eventually natural selection will wipe the altruists out. Even a very small number of egoists will produce this effect.

So far, this is just to repeat Williams's argument that indiscriminate altruism is an evolutionarily unstable strategy. However, a third type of entity was introduced to the simulation, one that followed the "tit-for-tat" strategy. As in the previous simulations, it was found that the tit-for-tatter fared better than either egoists or indiscriminate altruists, and moreover better than many other more elaborate strategies that were devised.

At first glance, this seems to just confirm what everybody already acknowledges: reciprocal altruism is an evolutionary stable strategy. However, there are some new consequences that are more encouraging for altruistophiles than that earlier message. The tit-for-tat strategy involves being helpful the first time you meet an entity, so that if the entity the tit-for-tatter meets is an altruist they will end up helping each other all the time. Likewise, if the entity they meet is another tit-for-tatter, they will end up helping each other all the time. So a population composed entirely of tit-for-tatters will be behaviourally indistinguishable from one composed entirely of indiscriminate altruists. Moreover, a predominance of tit-for-tatters in a population allows indiscriminate altruists to do reasonably well, albeit they will not do as well as the tit-for-tatters as long as there are egoists around. So altruists can remain as a good-sized minority.

In yet further variants, more complications were introduced in efforts to make the scenario more realistic. Among the complications introduced was making it possible for "mistakes" to be made. A random factor was introduced whereby an individual would occasionally defect in a situation where its strategy would dictate cooperation, or vice versa. This created problems for tit-for-tatters, because if two tit-for-tatters meet, and one mistakenly defects, they can end up defecting against each other forever. This version of the game was won by a strategy called "generous" or "tit-for-two-tats". On the tit-for-two-tats strategy, the entities help others they meet unless the others have defected in the last *two* encounters. Another alteration that was tried was alternating moves instead of moving at the same time; still another was making it possible to refuse to play in any given round – neither cooperating nor defecting. I will not discuss all the various complications here.[4] What Axelrod's game of life showed was that it was not simply inevitable that selfishness was the most rational strategy, even from the point of view of one's own selfish interests. Depending on how the game was set up, different strategies could win. One conclusion we can draw from this is that depending on how the environment of evolutionary adaptedness is set up, different strategies are evolutionarily stable. In some variants of the game, there was no clear winner. Instead, as one strategy became predominant, this made it advantageous to follow a different strategy, thus producing interesting cycles.

Despite all these complications, however, it turned out that many of the winners of different versions shared the tit-for-tatter's strategy of *default cooperation*, that is, they cooperated the first time they met any individual. The upshot of all this is that it is, at least in many scenarios, an evolutionary stable strategy to follow the maxim *"as a default*, do unto others as you would have them do unto you; but not if X, Y or Z"*. J. L. Mackie (1978; 1981) has argued that this is in fact how people usually behave anyway, and that evolutionary reasoning confirms the already reasonable view that this is better than being indiscriminately altruistic.

Many people feel, however, that this is a counsel of despair. It suggests that our innate dispositions are not to help someone unless there is an expectation of getting some benefit in return. This, in turn, sounds not like altruism but like self-interest: helping another as an indirect way of increasing the chances of reaping benefits for oneself. La Rochefoucauld summed up the thought: "When we help others in order to commit them to help us under similar circumstances,

[the] services we render them are, properly speaking, services we render to ourselves in advance" (quoted in Sesardic 1995: 130). A quick response to this is to return to the point I made earlier, that finding someone attractive is not the same as having the thought "it would enhance my fitness to have that person as a partner". We are liable to get confused about this because an answer to the question "Why do people do X?" can be "because it enhances their genetic fitness". But the only sense of "because" here that is warranted by evolutionary reasoning is the sense that that is the evolutionary explanation for why they do it, *not* that it is a person's motivation for doing it. Similarly, if I have an innate inclination to help others, and that inclination is there because it likely to bring some benefit to me, that does not mean that when I help others I am motivated by a desire to help myself. It could be that there is an innate inclination to help others who have helped me (or at least who have not harmed me) *full stop*. The distinction between an adaptive reason for something and a person's motivation for it is perfectly obvious in some cases, but seems to be forgotten in others. Clearly, I eat in order to stay alive. But that is true in the sense that the need to eat to stay alive is the adaptive explanation for why we eat, and have the disposition to eat. But I do not eat out of a desire to survive; I eat because I am hungry. David Buller makes this point very forcefully:

> We would never be tempted to think that livers contribute to success in the game of reproductive competition in virtue of having internalised adaptive goals; and the fact that psychological mechanisms themselves contain goals should make us no more tempted to suppose that the reason they contribute to reproductive success is that their internal goals are adaptive goals. (1999: 108)

In fact, we can push this thought a little further. We need not stop with just saying that, for all evolutionary reasoning tells us, we might not have a thought in our heads about getting a benefit when we help someone else. One of the evolutionary psychologist's arguments for the modularity thesis is that dedicated mechanisms are better at their tasks than more multi-purpose ones. If a situation demands a response immediately, it is better if that response is automatic, that is, done without weighing up the pros and cons first. Of course, we *do* often weigh up the pros and cons of helping someone, but it at least fits with evolutionary psychology's typical pattern of explanation to say that we also have spontaneous impulses to help people. Even if those impulses are only directed to our kin and those who have helped us (or, using tit-for-tat, those who we do not believe to have harmed us), that surely is better than nothing.

But it still might not be news as good as we would like. Even if the impulses to help others are spontaneous rather than calculated, we are still being discriminating in whom we help. Surely, we might want to say, moral obligations are *universal*, in the sense that we have duties to everyone, not duties that discriminate in the way that kin altruism and reciprocal altruism do. The upshot of the evolutionary arguments seems to be that we have no natural inclination to act on those duties. The conclusion seems to be that instead of having a natural inclination to treat everyone the same, we have a natural inclination to treat some people better than others.

But actually, the claim that we have universal duties is not the same as the claim that we should treat everyone the same. First, a plausible example of a moral universal is that anyone who commits a crime should be punished. This is universal in the sense that an "if ... then ..." statement applies to everyone: if you commit a crime you should be punished, no matter who you are. But it does not demand that we treat everyone the same; it demands that we treat criminals differently from others. The judicial system, we might say, practises tit-for-tat on our behalf.

Secondly, the evolutionary arguments tell us that we probably have an inclination to favour our kin. But, in fact, many of our laws enshrine the idea that it is reasonable to do so. Parents are required by law to provide for their children; if someone dies without leaving a will, their money passes to their next of kin; and so on. Admittedly, as I indicated in §13.1, evolutionary argument also suggests that parents will be inclined to favour strong or good-looking children over weak or ugly ones. Clearly, more needs to be said on this issue, which I will defer until the next section.

Finally, perhaps the opposition between acting for someone else's benefit and acting for your own is not a good way to capture the distinction between acting morally and not acting morally. It is tempting to think that we would need to explain how hardcore, non-kin altruism could evolve before we could explain how *morality* could evolve. But consider the following two scenarios: I am morally indignant because of the way I have been treated; and I am morally indignant because of the way someone else has been treated. No doubt, my indignation is more likely to be aroused when I am the victim than when someone else is. (Again, it did not need evolutionary psychology to tell us that.) No doubt, also, when I am the victim my protestations of moral indignation are sometimes a sham masking purely selfish motives. But neither of these means that if I am morally indignant on my own behalf it must be something other than genuine *moral* indignation. Let us say I have a moral sense that one ought not to treat people in certain ways. It is at least possible, and surely happens at least sometimes, that the moral sense that motivates me to protest at myself being treated that way is the same as the moral sense that motivates me to protest at someone else being treated that way. The feeling "that's not fair!" can be a very powerful one, and fairness is a moral concept. And the reaction "that's not fair!" can be called up either by something done to me, or something done to someone else. My point is that the question of whether people are innately egoistic or altruistic is not the same as the question of whether people are innately moral or not. As far as relevance to moral questions go, the whole debate about whether altruism is likely to have evolved in human beings may be a red herring.

14.5 Intuitions again

I am not, however, claiming that people are innately moral; people have tendencies to be selfish, lazy, violent, unfair and much more besides. Only the most naive 1960s anthropologists think that these unpleasant characteristics are entirely due to upbringing. Evolutionary psychology has some bad news (or insidious propaganda, depending on your point of view) on the moral front. It tells us that, even if many men manage to restrain themselves from philandering, they are likely to be often *tempted* to philander. It also tells us that we are disposed to favour our kin, which does not just lead to parental care, but also to nepotism. It also tells us that parents are likely to be tempted to favour their strong, good-looking children over weak, ugly ones. Even if the tit-for-tat strategy has evolved in human beings, that may be a mixed blessing. We are familiar with the expression "tit-for-tat killings".

Despite all this, evolutionary psychology has some good news too. Evolutionary psychologists tell us that we have a cognitive architecture that is rich in concepts and is universal. To reiterate what I wrote in Chapter 13, this does not mean that everyone has it or that it is inevitable, but that it transcends cultures and is highly likely to develop despite the vicissitudes of the environment. This does not just include the nasty things in life, but also includes friendship, love, intellectual curiosity and the impulse to make and appreciate art, stories, music and jokes. Moreover, I want to suggest that it may include moral perceptions as well.

A moral perception is distinct from a desire. It is coherent to have the thought "I do not want to do X, but I ought to do X". For that to be a *moral* perception, it cannot mean the same thing as "In one way I do not want to do X, but in another way I do want to do X". I shall go out on a limb here and suggest that we all in fact recognize this difference, even if some moral philosophers have tried to deny it. The perception that certain things are right or wrong may, I suggest, be part of our evolutionary heritage as well.

Recall what I suggested in §14.4: we are familiar with the reaction "that's not fair!" Children develop this reaction, it seems, very young. Parents may often have to tell a child that something is not fair, but that does not mean that they have to explain to the child what fairness is. It may be that this reaction is a part of our evolutionary heritage. There is evidence that other primates react strongly against situations that we would call unfair. For example, studies have been carried out with capuchin monkeys, wherein they were initially trained to give tokens to human experimenters, and received a piece of cucumber as a reward (Brosnan & de Waal 2003; Brosnan 2006). The scenario was then changed, with some monkeys receiving a grape (which they apparently value more highly than a piece of cucumber) instead of a cucumber, in exchange for the same token. The ones who received what they perceive as the "lesser" reward were rather unhappy about this, often rejecting the piece of cucumber – for example, throwing it away – even though they had been perfectly happy with a piece of cucumber before. So it seems that their willingness to accept the situation depends on how they see other individuals being treated. In other words, they are unhappy with unequal treatment. As we might put it, they do not like being treated unfairly. Capuchin monkeys are social creatures, among whom sharing food is common. What these studies suggest is that a sensitivity to fairness and unfairness is an adaptation to social living, and one that predates the emergence of human beings. Similar behaviour has been found in chimpanzees.

To repeat, we may often in practice be more sensitive to instances where we ourselves are the victim. (The capuchin monkeys in fact are.) But we understand the concept just as well when it apples to another, and indeed the very idea of fairness entails that the same rules apply to me as to other people. I will admit, as well, that different cultures apparently had, and have, no problem doing things that we here and now think are grossly unfair. The Spartans apparently had no problem dumping weak babies on mountainsides. But that does not prove that they lacked the concept of fairness. They may have thought that the rule "you should be left on a mountainside to die if you are weak" was a universal rule, in the same way that we think that "you should be put in prison if you have murdered" is. Perhaps the concept of fairness is part of our evolved cognitive apparatus, which can be harnessed just as various evolved cognitive mechanisms are harnessed when we learn to drive a car or programme a VCR, or when scientists split the atom. At one time, few people thought it unfair that women could not vote or go to university, but the writings of Mary Wollstonecraft and J. S. Mill helped people to see these things as unfair.

The concept of fairness is, I suggest, an intuition, in the way that I earlier suggested that the concept of treating people as ends is an intuition. In both cases, if someone showed no signs of possessing the concept we would find that person morally lacking. And there are probably many others too. I have been using the concept "intuition" rather a lot in this chapter. As I assume is clear by now, what I want to capture with the word is an unreasoned sense that, as it were, just "hits you" that something is good or bad, right or wrong. I take it that this experience is familiar to readers of this book. I would claim, further, that at least some of these specifically moral intuitions are human universals: "murder is wrong" is a plausible candidate. One might say that we *just see* that murder is wrong. If someone does not see it no amount of reasoning will show them, but there is usually no need to try, any more than we need to reason with someone to convince them that sugar is sweet. There are people who apparently do not see that murder is wrong, and they are called psychopaths. Sometimes they say that they *know* it is wrong, but this knowing, whatever it is, is not the knowing that most people have that murder is wrong.[5] But a perception that such-and-such is such-and-such that just "hits you", is unreasoned and is a human universal, is exactly the kind of thing that evolutionary psychologists say that we human beings have as part of our evolutionary heritage. You do not *infer* that someone is attractive, or that solid objects cannot pass through other solid objects. So evolutionary psychology makes more plausible the claim that there are humanly universal moral intuitions.

In practice, one often finds moral philosophers attempting to show that their own view fits with our intuitions and that their opponents' views do not. For example, the literature on utilitarianism is full of anti-utilitarians attempting to show that utilitarianism violates some moral intuition, and utilitarians trying to show that it does not. But if intuitions are regularly used in this way, then it seems that they are being given considerable weight. But two more things can be said about intuitions.

First, it is not clear, at least to me, how moral arguments could ever get going in the first place, if there were no moral intuitions. How could we convince someone that something was wrong, except by showing that it is in some way *the same kind of thing* as something they already accept is wrong? How could moral philosophers argue for some general moral principle – such as "maximize happiness" – except by showing that it fits reasonably well with what we already think is right and what wrong?

Secondly, if we really do have moral intuitions that are part of our evolutionary heritage, in the way that I have suggested the concept of fairness is, then they probably run very deep. That means that anyone trying to flout them, or trying to get us to flout them, in a large-scale way is *going against the grain*, in the way that Stalin's attempts to abolish romantic love in the 1930s went against the grain. That, in itself, is a reason to respect them.

Regardless of the empirical specifics, evolutionary psychology gives us good reason in general to believe that there are moral intuitions. Can it also tell us what our intuitions are? Maybe. But, as I argued above, it is not clear why it would be better than just consulting our intuitions. Surely, if an intuition is strong enough a scientific argument that told us that we have it would not make any difference. But the theory of evolution does have one more contribution to make. The frame problem, and the evolutionary arguments that back it up give us general reasons to think that we have *many different intuitions*. This means that any ethical theory that has one single principle – for example utilitarianism – is unlikely to fit them all.

Notes

1. The argument in Darwin's *Origin*

1. Evolutionists often associate the term "essentialism" with the names of Plato and Aristotle. For example, "Darwin undid the essentialism that Western philosophy had inherited from Plato and Aristotle, and put variation in its place" (D. J. Futuyma, *Evolutionary Biology*, 3rd edn [Sunderland, MA: Sinauer, 1998], 8). Whether the doctrine, as presented by people like Dennett, is an accurate reflection of either Plato's or Aristotle's actual views is not a question I shall address here.
2. I shall take up the issue of religion in Chapter 12, and the issue of essentialism in Chapter 9.
3. Richard Dawkins disapprovingly reports that a "distinguished modern philosopher", whom he does not name, once suggested this to him (*The Blind Watchmaker* [Harmondsworth: Penguin, 1986], 5).
4. Some theorists (e.g. W. Wimsatt, "Generativity, Entrenchment, Evolution, Innateness", in *Where Biology Meets Psychology*, V. G. Hardcastle [ed.], 139–80 [Cambridge, MA: MIT Press, 1999]) have argued for a revival of what they call Lamarckian ideas. But the "Lamarckism" that they defend consists of the claim that "inheritance of acquired traits" exists, rather than that it is the primary explanation for good design. So it does not conflict with Darwin's theory.
5. A pigeon-fancier is not just someone who is fond of pigeons, or who keeps pigeons, but specifically someone who *breeds* pigeons to produce unusual ("fancy") traits.
6. The historian Eric Hobsbawm holds that Malthus's empirical claims have been decisively refuted by what has actually happened in the world since 1945: food production has in fact increased faster than population, largely due to advances in agricultural technology (E. Hobsbawm, *Age of Extremes: The Short Twentieth Century 1914–1991* [London: Abacus, 1994], 260). The real problem is with how that food gets distributed, with large numbers of people in the wealthy countries eating not just more than they need, but more than is good for them. Nonetheless, the problem of human population growth is a troubling one, and is discussed in chilling terms by Colin Tudge, *The Variety of Life: A Survey and Celebration of All the Creatures that have Ever Lived* (Oxford: Oxford University Press, 2000), 609ff.
7. Malthus's views were a precursor of the "social Darwinism" of the late-nineteenth and early-twentieth centuries – an instance of an attempt to derive ethical prescriptions from the theory of evolution. I shall return to this issue in Chapter 14.

8. Darwin follows this quote with more specific examples.
9. "*Ceteris paribus*" means "all other things being equal". It draws attention to the fact that biology is a science without strict laws of its own. This raises issues that will be discussed in Chapter 10.
10. The much-used example of the black and white pepper moths is not sufficient, for that is an instance of changes in the relative frequency of already existing traits, not of the production of novel traits.
11. See, for example, R. Dawkins, "Universal Darwinism", in *Evolution from Molecules to Man*, D. S. Bendall (ed.), 403–25 (Cambridge: Cambridge University Press, 1983), which I shall discuss in Chapter 4.

2. The power of genes

1. I have borrowed the metaphor of "pillars" from Stephen Jay Gould, *The Structure of Evolutionary Theory* (Cambridge, MA: Harvard University Press, 2002). On Gould's account, what he (slightly misleadingly) calls the "modern synthesis" has *three* pillars, but I do not find his division helpful in terms of expository clarity.
2. For example, see S. J. Gould & R. Lewontin, "The Spandrels of San Marco and the Panglossian Paradigm: A Critique of the Adaptationist Programme", *Proceedings of the Royal Society of London* B205 (1979), 581–98. I shall discuss this famous article at length in Chapter 4.
3. See, for example, R. Dawkins, "In Defence of Selfish Genes", *Philosophy* 56 (1979), 556–73, which is a response to M. Midgley, "Gene-Juggling", *Philosophy* 54 (1979), 439–58.
4. If gene-centrist evolutionists read Aristotle, they might say that genes are the *formal cause* of an organism. As we shall see shortly, they see genes as the final cause of an organism as well.
5. Although Colin McGinn comes pretty close; see his *The Mysterious Flame: Conscious Minds in a Material World* (New York: Basic Books, 1999), 224ff.
6. The question of whether, and how far, this is true is at the heart of the heated battles over sociobiology and evolutionary psychology, which will be the subjects of Chapter 13.

3. Units of selection

1. All of these statements are *ceteris paribus*. As I noted earlier, this is a pervasive feature of biological "laws" that you just have to get used to. See Chapter 10.
2. Or very nearly random; see the discussion of segregation-distorter genes below.
3. Actually I have simplified things a little here. There is no one gene that determines eye colour, but brown eyes and blue eyes do generally show a dominant–recessive pattern.
4. See S. Okasha, "Why Won't the Group Selection Controversy Go Away?", *British Journal for the Philosophy of Science* 52 (2001), 25–50, for a commentary on this debate.
5. Bill Drummond of the KLF wrote a book entitled *The Manual: How to Have a Number One the Easy Way* (KLF Publications, 1988). This book was read by the Austrian group Edelweiss who, by following Drummond's recommendations, produced a single that reached number one all over Europe.
6. See, for example, Dawkins's provocatively titled paper "Viruses of the Mind", in *Dennett and his Critics*, B. Dahlbom (ed.), 13–27 (Oxford: Blackwell, 1993).
7. In fact, Sterelny has changed his position since the time he wrote the paper with Kitcher that I discussed above, and is now essentially a multi-level selectionist. See K. Sterelny, K. C. Smith, M. Dickison, "The Extended Replicator", *Biology and Philosophy* 11(3)

(1996), 377–403. Moreover, Dawkins is strictly speaking a multi-level selectionist, since he believes in memes as entities that replicate and evolve independently of genes.

4. Panglossianism and its discontents

1. Or at least at some microscopic level, whatever the equivalent of "genes" is on an alien planet.
2. There has been some excitement recently around "complexity theory", a mathematical model that is purported to explain the spontaneous emergence of complexity; see S. Kaufmann, *At Home in the Universe: The Search for the Laws of Self-Organization and Complexity* (Oxford: Oxford University Press, 1995). It remains to be seen, however, whether it can explain specifically *functional* complexity in a way that makes natural selection redundant.
3. Jerry Fodor, *The Mind Doesn't Work That Way* (Cambridge, MA: MIT Press, 2000) argues a similar point.
4. Gould uses this expression frequently in *The Structure of Evolutionary Theory*.
5. A similar point is made by R. Lewontin, *The Triple Helix: Gene, Organism and Environment* (Cambridge, MA: Harvard University Press, 2000).
6. Evolutionary psychologists envisage their project as including the use of adaptationism in this way. This is a feature of evolutionary psychology that has drawn sceptical comment from P. Sheldon Davies, "The Conflict of Evolutionary Psychology", in *Where Biology Meets Psychology*, V. G. Hardcastle (ed.), 67–82, and T. Grantham & S. Nichols, "Evolutionary Psychology: Ultimate Explanations and Panglossian Predictions", in *Where Biology Meets Psychology*, V. G. Hardcastle (ed.), 47–66.

5. The role of development

1. Darwin wrote in the margins of his copy of Chambers's *Vestiges of Creation*: "Never say higher or lower in referring to organisms". Gregory Radick, "Two Explanations of Evolutionary Progress", *Biology and Philosophy* 15 (2000), 475–91, discusses the meaning of this.
2. An excellent collection of pictures of Burgess Shale fossils can be found in D. E. G. Briggs, D. H. Erwin, F. J. Collier, C. Clark, *The Fossils of the Burgess Shale* (Washington DC: Smithsonian Institution Press, 1994).
3. For two views on this debate, see S. J. Gould, *Wonderful Life* (London: Vintage, 2000) and R. Dawkins, "The Velvet Worm's Tale", in his *The Ancestor's Tale: A Pilgrimage to the Dawn of Life*, 362–79 (London: Weidenfeld & Nicolson, 2004).
4. See S. B. Carroll, *Endless Forms Most Beautiful: The New Science of Evo-Devo* (New York: Norton, 2005), esp. ch. 3, for a fuller account of these discoveries.
5. See C. Stern (ed.), *Gastrulation: From Cells to Embryo* (Cold Spring Harbor, NY: Cold Spring Harbor Laboratory Press, 2004) if you want to know all the intricate details of this process.

7. Function: "what it is for" versus "what it does"

1. For some good discussions on anthropomorphizing see S. Glendinning, "Heidegger and the Question of Animality", *International Journal of Philosophical Studies* 4(1) (1996), 67–86 and B. Keeley, "Anthropomorphism, Primatomorphism, Mammalomorphism: Understanding Cross-Species Comparisons", *Biology and Philosophy* 19(4) (2004), 521–40.
2. This was Beckner's second stab at a definition.

8. Biological categories

1. This example comes from I. M. Copi, "Essence and Accident", *Journal of Philosophy* 51(23) (1954), 706–19, esp. 710. His point, more specifically, was to show that what we consider an essential change, and what an accidental change, is contingent on our language. I am grateful to Alasdair Richmond for helping me to locate the source.
2. See G. G. Simpson, *Principles of Animal Taxonomy* (New York: Columbia University Press, 1961) for a presentation of the evolutionary approach.
3. See M. Ereshefsky, *The Poverty of the Linnaean Hierarchy: A Philosophical Study of Biological Taxonomy* (Cambridge: Cambridge University Press, 2000) for a discussion of this issue.

9. Species and their special problems

1. This is an instance of what philosophers call a Sorites paradox.
2. See D. A. Baum and M. Donoghue, "Choosing Among Alternative 'Phylogenetic' Species Concepts", in *Conceptual Issues in Evolutionary Biology*, 3rd edn, E. Sober (ed.), 387–406 (Cambridge, MA: MIT Press, [1995] 2006) for an overview of alternative criteria for specieshood.

10. Biology and philosophy of science

1. I write "entity" rather than "organism" to remain neutral on the units of selection question. You may, however, find it helpful to think about organisms.
2. Strictly speaking, Newton's equations *always* yield results at variance with those yielded by the equations of the theory of relativity, but at sub-light speeds these differences are so small as to be negligible. This complication need not concern us here. Neither need we be concerned with the question of the truth-value of superseded theories, the question that motivates "approximate realism".
3. Not everyone thinks this story fits physics, in fact. See N. Cartwright, *The Dappled World: A Study of the Boundaries of Science* (Cambridge: Cambridge University Press, 1999).

11. Evolution and epistemology

1. See K. Popper, *Objective Knowledge: An Evolutionary Approach*, 2nd edn (Oxford: Oxford University Press, 1979), ch. 1 for Popper's clearest exposition of the problem of induction and the falsificationist criterion.

12. Evolution and religion

1. See J. Leslie, *Universes* (London: Routledge, 1989) for an impressive list of examples of initial conditions for life.
2. An example of a computer-generated simulation of evolution is John Conway's "Game of Life"; for a brief description see, D. Dennett, *Darwin's Dangerous Idea: Evolution and the Meanings of Life* (Harmondsworth: Penguin, 1995), 166–76.
3. Unless, that is, the "laws" of biology were themselves the fundamental laws. But in a world where that was the case, the capacity of the theory of evolution to explain the bodily processes of organisms in terms of mechanical forces would be lost.
4. This trivial point is sometimes called the *anthropic principle* but that name is given to a number of different principles, some of them highly controversial.

5. If this were really the case, we would have no rational assurance that any of the laws currently in operation at present would continue to operate in the future, even five seconds in the future. But Hume believed we had no rational assurance of this anyway.

6. I shall give just one example here:

> Not to be born at all is best, far best that can befall; next best, when born, with least delay to trace the backward way. For when youth passes with its giddy train, troubles on troubles follow, toils on toils. Pain, pain for ever pain; and none escapes life's coils. Envy, sedition, strife, carnage and war, make up the tale of life. (Sophocles: *Oedipus at Colonus*, F. Storr (trans.)
> [London: Heinemann, 1912])

13. Evolution and human nature

1. Morris's book contains what appear today to be serious methodological errors. For example, at the outset he announces that his focus is on "the ordinary, successful members of the major cultures", and dismisses "simple tribal groups" as "stultified" (D. Morris, *The Naked Ape* [London: Jonathan Cape, 1967], 10). But if one's aim is to discover traits that are a legacy of our evolutionary past, then surely it is important to study cultures that have had little contact with the outside world.

2. See U. Segerstråle, *Defenders of the Truth: The Sociobiology Debate* (Oxford: Oxford University Press, 2000), 22ff., for a fuller description of this and similar incidents. For Wilson's own thoughts on these incidents, see E. O. Wilson, "Science and Ideology", *Academic Questions* 9(4) (1995), 73–81.

3. See C. Badcock, *Evolutionary Psychology: A Critical Introduction* (Cambridge: Polity, 2000), 106–10 for a brief summary of criticisms of Cosmides' interpretation of the Wason selection task experiments.

4. Both of these examples come from Morris, *The Naked Ape*.

14. Biology and ethics

1. For criticisms of evolutionary psychology along these lines, see most of the contributions to H. Rose & S. Rose (eds), *Alas, Poor Darwin: Arguments Against Evolutionary Psychology* (London: Jonathan Cape, 2000).

2. This variant of the prisoner's dilemma comes from D. Hofstadter, *Metamagical Themas: Questing for the Essence of Mind and Pattern* (New York: Basic Books, 1985), ch. 29.

3. For fuller, and very lucid, accounts of all this, see *ibid.*, and M. Ridley, *The Origins of Virtue* (Harmondsworth: Penguin, 1996), ch. 3.

4. For discussions of game of life complications see Hofstadter, *Metamagical Themas*, and Ridley, *The Origins of Virtue*, ch. 4.

5. See W. Glannon, "Psychopathy and Responsibility", *Journal of Applied Philosophy* 14(3) (1997), 263–75, for a discussion of this.

Further reading

Full bibliographic details are given in the bibliography.

General

There are a number of very good introductions to the theory of evolution. John Maynard Smith's *The Theory of Evolution* (1993) and Ernst Mayr's *One Long Argument* (1993) can both be highly recommended. More in-depth (and much longer) presentations of evolution are *Evolutionary Biology* by Douglas Futuyma (1998), and *Evolution*, edited by Peter Skelton (1993). The latter two may be heavy going for non-biologists, but are well worth it for the wealth of fascinating information about living creatures that they contain. The same can be said for Colin Tudge's *The Variety of Life* (2000), which is not a book about evolution but, as the subtitle says, "a survey and celebration of all the creatures that have ever lived". Richard Dawkins is the master at conveying the excitement and wonder of the study of living things. Particularly good in this regard is his *The Ancestor's Tale* (2004), which, apart from tracing our evolutionary past, contains numerous interesting theoretical *aperçus*. Daniel Dennett's *Darwin's Dangerous Idea* (1995) shows how the theory of evolution radically affects the way we think about a host of issues, although it is rather partisan on some topics.

The anthologies *The Philosophy of Biology*, edited by David Hull and Michael Ruse (1998), and *Conceptual Issues in Evolutionary Biology*, edited by Elliott Sober (1994; 2006), are extremely useful, and between them contain many of the classic papers in the area. Although it is less philosophical in orientation, *Evolution*, edited by Mark Ridley (1997), can also be recommended.

1. The argument in Darwin's *Origin*

The Penguin Classics edition of Darwin's *The Origin of Species* ([1859] 1968) is a reprint of the first edition of 1859, and is the one to read. Later editions were somewhat watered down by Darwin in light of contemporary criticisms that, in retrospect, fail to hold water. The Penguin edition also includes Darwin's Historical Sketch and a glossary. Ruse's *The Darwinian Revolution* (1999) presents a very good analysis of Darwin's arguments, and sets them in their historical context. Part I of *The Cambridge Companion to Darwin*, edited by Jonathan Hodge and Gregory Radick (2003) consists of essays on Darwin's theories, while Part II deals with the historical context. Another useful discussion of Darwin's theories can be found in Tim Lewens's *Darwin* (2006).

2. The power of genes

The most forceful presentation of the gene-centred view of evolution is Dawkins's *The Selfish Gene* ([1976] 1989). This work explains such crucial concepts as Hamilton's rule, the difference between a gene's-eye view and an organism's-eye view, and the implications of the gene-centred view for cooperation and altruism in a most accessible and exciting style. The classic works that laid the foundations of this view are George Williams's *Adaptation and Natural Selection* (1966) and William Hamilton's "The Genetical Evolution of Social Behaviour" (1964a,b).

3. Units of selection

Sober defends individual-level selection in *The Nature of Selection* (1984), and, with David Sloan Wilson, defends group selection in *Unto Others* (1998). Further discussion of group selection can be found in Samir Okasha's "Why Won't the Group Selection Controversy Go Away?" (2001). A small but useful collection of essays on units of selection can be found in Part III of Hull and Ruse's *The Philosophy of Biology* (1998). This includes Kim Sterelny and Philip Kitcher's defence of "pluralist gene-selectionism".

Darwinizing Culture, edited by Robert Aunger (2001) is a collection of essays on memes. A lively and accessible defence of memes is Susan Blackmore's *The Meme Machine* (1999).

4. Panglossianism and its discontents

Dawkins's "Universal Darwinism" (1983) can also be found in Hull and Ruse's *The Philosophy of Biology* (1998). To avoid getting an exaggerated idea of how extreme Dawkins's position is, readers should also look at his *The Extended Phenotype* (1982), Chapter 3, "Constraints on Perfection". Stephen Jay Gould and Richard Lewontin's classic paper, "The Spandrels of San Marco and the Panglossian Paradigm" (1979), can also be found in Sober's *Conceptual Issues in Evolutionary Biology* (1994; 2006). Niles Eldredge and Gould's (1972) account of "punctuated equilibria" can be found in their paper of that title. Gould and Elizabeth Vrba's (1982) account of "exaptation" can be found in Hull and Ruse's *The Philosophy of Biology* (1998). Dennett responds to the accusation of "panglossianism" in "Intentional Systems in Cognitive Ethology: The 'Panglossian Paradigm' Defended" (1987). More technical discussions of the nature of adaptation can be found in *The Latest on the Best*, edited by John Dupré (1987), and *Adaptationism and Optimality*, edited by Steven Hecht Orzack and Sober (2001).

A brief and extremely accessible overview of the disputes between Dawkins and Gould is Sterelny's *Dawkins vs. Gould* (2001). An amusing, brief discussion of Gould and Lewontin is David Queller's "The Spaniels of St Marx and the Panglossian Paradox" (1995).

5. The role of development

Sean Carroll's *Endless Forms Most Beautiful* (2005) gives an accessible account of evo-devo and why it is important. Jeffrey Schwartz's *Sudden Origins* (1999) gives a good account of the discovery of homeobox genes and the impact that has had on evolutionary theory. Scott Gilbert's *Developmental Biology* (2003) is a thorough textbook on the subject, but it may be a bit difficult for non-biologists. Gould's *Ontogeny and Phylogeny* (1977) gives an account of the nineteenth-century theories that in some ways anticipated the insights of evo-devo, and was itself a prescient forerunner of the movement. Readers with plenty of time on their hands could try Gould's enormous book *The Structure of Evolutionary Theory* (2002), wherein he presents his fullest account of how evolutionary theory has changed since the "modern synthesis" of the mid-twentieth century.

A clear and succinct statement of the principles of developmental systems theory can be found in Paul Griffiths and Russell Gray's paper "Developmental Systems and Evolutionary Explanation" (1994), which is in Hull and Ruse's *The Philosophy of Biology* (1998). A fuller presentation and defence is Susan Oyama's *The Ontogeny of Information* (1985). A collection by Oyama, Griffiths and Gray called *Cycles of Contingency* (2001) includes discussions of the theory's implications, plus a number of criticisms of the theory.

6. Nature and nurture

Stephen Stich's *Innate Ideas* (1975) is an anthology of classic and (then) contemporary writings on the subject, and its Introduction gives Stich's own account of what "innate" means. Lewontin's critique of the idea that organisms "develop" from their genetic codes can be found in his *The Triple Helix* (2000a), Chapter 1. The interactionist consensus is ably explained and defended by Ridley in *Nature via Nurture* (2003b). André Ariew's "Innateness is Canalization" and William Wimsatt's "Generativity, Entrenchment, Evolution, Innateness" can be found in *Where Biology Meets Psychology*, edited by Valerie Gray Hardcastle (1999). James Maclaurin's "The Resurrection of Innateness" (2002) and Griffiths's "What is Innateness?" (2002) both appeared in *The Monist* 85(1).

7. Function: "what it is for" versus "what it does"

The collection *Nature's Purposes*, edited by Colin Allen *et al.* (1998), contains classic papers on function – Robert Cummins, "Functional Analysis", Ruth Millikan, "In Defense of Proper Functions" and Larry Wright, "Functions" – as well as numerous other useful contributions by both philosophers and biologists. A special edition of the journal *Studies in History and Philosophy of Science*, part C, 31(1) (2000) was devoted to papers on function. The parallels between function in biology and in human design are explored in Lewens's *Organisms and Artifacts* (2005).

8. Biological categories

A forceful defence of the view that cladistics is the only non-arbitrary way to classify organisms is presented by Dawkins in *The Blind Watchmaker* (1986) in the chapter "One True Tree of Life". The same idea is discussed at greater length in Mark Ridley's *Evolution and Classification* (1989). Ridley also provides a helpful overview of the issue in *Evolution* (1997; 2003a), ch. 16. The section called "Systematic Philosophies" in Sober's *Conceptual Issues in Evolutionary Biology* (2nd edn, 1994) contains some important papers on the topic. A pluralistic approach to taxonomy is defended by Marc Ereshefsky in his *The Poverty of the Linnaean Hierarchy* (2000).

9. Species and their special problems

Ernst Mayr's *Toward a New Philosophy of Biology* (1988), Chapter 6, presents the issue of species in an engaging and accessible way. The same can be said for "The Salamander's Tale" in Dawkins's *The Ancestor's Tale* (2004), which provides a striking example of a "ring species". Some important papers on the topic are to be found in Hull and Ruse's *The Philosophy of*

Biology (1998), Part V. The collection *Species*, edited by Robert Wilson (1999), presents the issue from a number of different points of view. An older collection of important papers on the topic is *The Units of Evolution*, edited by Ereshefsky (1992).

10. Biology and philosophy of science

On the issue of laws, or the lack thereof, in biology, see Alex Rosenberg's *Instrumental Biology or the Disunity of Science* (1994), John Beatty's "The Evolutionary Contingency Thesis" ([1995] 2006), and Sober's "Two Recent Outbreaks of Lawlessness in Recent Philosophy of Biology" in *Conceptual Issues in Evolutionary Biology* (3rd edn, 2006).

The classic defence of a pluralistic view of science is Paul Feyerabend's *Against Method* (1993). More recent presentations of this view (not quite as extreme as Feyerabend's) are Nancy Cartwright's *The Dappled World* (1999) and Dupré's *The Disorder of Things* (1993). The latter, in particular, makes much of the ways in which biology is unlike physics and chemistry.

11. Evolution and epistemology

Karl Popper's "evolutionary" approach to knowledge is outlined in his *Objective Knowledge* (1979). W. V. Quine's epistemology is usefully summarized in his *From Stimulus to Science* (1995). The collection *Issues in Evolutionary Epistemology*, edited by Kai Hahlweg and C. A. Hooker (1989), provides an excellent overview of the subject.

12. Evolution and religion

Dawkins's *The Blind Watchmaker* (1986) is a classic presentation of the argument that evolution dispenses with the need to posit an intelligent designer. The literature on the argument from design is vast. For a very good overview of recent thinking on the subject, see Neil Manson's collection *God and Design* (2003), which includes arguments on both sides concerning both biological "intelligent design" and cosmological "fine-tuning" as evidence for the existence of God. It contains contributions by John Leslie and Michael Behe, as well as critical responses to them. More general discussions of the relationship between evolution and religion are Kenneth Miller's *Finding Darwin's God* (1999) and Michael Ruse's *Can a Darwinian Be a Christian?* (2004). An evolutionary psychological account of the origins of religion is given by Pascal Boyer in *Religion Explained* (2001).

13. Evolution and human nature

The final chapter of Edward O. Wilson's *Sociobiology* (1975) was what initially brought all the wrath down on his head. His *On Human Nature* (1978) gives his more considered views on the subject. Ulicia Segerstråle's *Defenders of the Truth* (2000) is an entertaining account of the controversies surrounding sociobiology, although it is a little biased against Lewontin.

John Tooby and Leda Cosmides's canonical statement of their position is "The Psychological Foundations of Culture" (1992), which is in *The Adapted Mind*, edited by Jerome Barkow, Cosmides and Tooby. That book as a whole can be recommended for presenting the views of orthodox evolutionary psychologists on mating, colour-perception, landscape preference and other topics. A more succinct statement of the basic tenets of evolutionary psychology is Cosmides and Tooby's internet article "Evolutionary Psychology: A Primer" (1997). A lively popular presentation of the field is Steven Pinker's *How the Mind Works* (1997). David Buss's *Evolutionary Psychology* (1999) is a textbook-style overview, but still very accessible.

There is a large volume of anti-evolutionary psychology literature, much of which is rather poor. Many of these naysayers fail to distinguish evolutionary psychology from sociobiology, and accuse evolutionary psychologists of genetic determinism. More informed critiques can be found in Christopher Badcock's *Evolutionary Psychology* (2000) and David Buller's *Adapting Minds* (2005). A rather more technical critique is Jerry Fodor's *The Mind Doesn't Work That Way* (2000). However, this book presupposes a good knowledge of contemporary philosophy of mind. Janet Radcliffe Richards' *Human Nature after Darwin* (2000) does an excellent job of defusing many of the worries that people have about evolutionary psychology's alleged social, political and moral implications.

14. Biology and ethics

Neven Sesardic's "Recent Work on Human Altruism and Evolution" (1995) is a useful introduction to the topic. *Issues in Evolutionary Ethics*, edited by Paul Thompson (1995), collects together a number of important articles on the subject, including ones by E. O. Wilson, Ruse, and William Rottschaefer and David Martinsen. A good overview of how evolution may account for moral beliefs and sentiments is Ridley's *The Origins of Virtue* (1996). This book is admirably clear, and covers reciprocal altruism, the prisoner's dilemma, tit-for-tat and other topics. Similar ground is covered in Brian Skyrms' *Evolution of the Social Contract* (1996). The classic work on these topics is Robert Axelrod's *The Evolution of Co-operation* (1990).

Bibliography

Allen, C., M. Bekoff, G. Lauder (eds) 1998. *Nature's Purposes: Analyses of Function and Design in Biology*. Cambridge, MA: MIT Press.

Allen, E., J. Beckwith, B. Beckwith *et al.* 1975. "Against 'Sociobiology'", letter in response to "Mindless Societies". *New York Review of Books* **22**(18) (13 November).

Ariew, A. 1999. "Innateness is Canalization". See Hardcastle (1999), 117–38.

Aristotle 2004. *On the Parts of Animals*, W. Ogle (trans.). Whitefish, MT: Kessinger.

Aunger, R. (ed.) 2001. *Darwinizing Culture: The Status of Memetics as a Science*. Oxford: Oxford University Press.

Axelrod, R. 1990. *The Evolution of Cooperation*, 2nd edn. Harmondsworth: Penguin.

Badcock, C. 2000. *Evolutionary Psychology: A Critical Introduction*. Cambridge: Polity.

Barkow, J., L. Cosmides, J. Tooby (eds) 1992. *The Adapted Mind*. Oxford: Oxford University Press.

Baum, D. A. & M. Donoghue [1995] 2006. "Choosing Among Alternative 'Phylogenetic' Species Concepts". In *Conceptual Issues in Evolutionary Biology*, 3rd edn, E. Sober (ed.), 387–406. Cambridge, MA: MIT Press.

Beatty, J. H. [1995] 2006. "The Evolutionary Contingency Thesis". Reprinted in *Conceptual Issues in Evolutionary Biology*, 3rd edn, E. Sober (ed.), 217–48. Cambridge, MA: MIT Press.

Beckner, M. 1969. "Function and Teleology". *Journal of the History of Biology* **2**: 151–64.

Behe, M. 1996. *Darwin's Black Box*. New York: Free Press.

Blackmore, S. 1999. *The Meme Machine*. Oxford: Oxford University Press.

Boorse, C. 1976. "Wright on Functions". *Philosophical Review* **85**: 70–86.

Boyer, P. 2001. *Religion Explained: The Human Instincts that Fashion Gods, Spirits and Ancestors*. London: Heinemann.

Brandon, R. 1988. "The Levels of Selection: A Hierarchy of Interactors". In *The Role of Behaviour in Evolution*, H. Plotkin (ed.), 51–71. Cambridge, MA: MIT Press.

Brandon, R. 2005. "Evolutionary Modules: Conceptual Analyses and Empirical Hypotheses". In *Modularity: Understanding the Development and Evolution of Natural Complex Systems*, W. Callebaut & D. Rasskin-Gutman (eds), 51–60. Cambridge, MA: MIT Press.

Briggs, D. E. G., D. H. Erwin, F. J. Collier, C. Clark 1994. *The Fossils of the Burgess Shale*. Washington DC: Smithsonian Institution Press.

Brosnan, S. F. 2006. "Nonhuman Species' Reactions to Inequity and their Implications for Fairness". *Social Justice Research* **19**(2): 153–85.

Brosnan, S. F. & F. B. M. de Waal 2003. "Monkeys Reject Unequal Pay". *Nature* **425**: 297–9.

Buller, D. 1999. "DeFreuding Evolutionary Psychology: Adaptation and Human Motivation". See Hardcastle (1999), 99–114.

Buller, D. 2005. *Adapting Minds: Evolutionary Psychology and the Persistent Quest for Human Nature*. Cambridge, MA: MIT Press.

Buss, D. 1999. *Evolutionary Psychology: The New Science of the Mind*. Boston, MA: Allyn & Bacon.

Buss, D. (ed.) 2005. *The Handbook of Evolutionary Psychology*. Hoboken, NJ: John Wiley.

Canfield, J. 1964. "Teleological Explanations in Biology". *British Journal for the Philosophy of Science* **14**: 285–95.

Carrier, M. 1995. "Evolutionary Change and Lawlikeness". In *Concepts, Theories and Rationality in the Biological Sciences: The Second Pittsburgh-Konstanz Colloquium in the Philosophy of Science*, G. Wolters & J. Lennox (eds), 83–97. Pittsburgh, PA: University of Pittsburgh Press.

Carroll, S. B. 2005. *Endless Forms Most Beautiful: The New Science of Evo-Devo*. New York: Norton.

Cartwright, N. 1999. *The Dappled World: A Study of the Boundaries of Science*. Cambridge: Cambridge University Press.

Chesterton, G. K. [1908] 2004. "Science and Religion". In *All Things Considered*, 77–80 Whitefish, MT: Kessinger.

Collins, A. G. n.d. *Introduction to Placozoa: The Most Simple of All Known Animals*. www.ucmp. berkeley.edu/phyla/placozoa/placozoa.html (accessed July 2007). University of California Museum of Paleontology.

Colyvan, M. & L. R. Ginzburg 2003. "Laws of Nature and Laws of Ecology". *Oikos* **101**(3): 649–53.

Copi, I. M. 1954. "Essence and Accident". *Journal of Philosophy* **51**(23): 706–19.

Cosmides, L. & J. Tooby 1992. "Cognitive Adaptations for Social Exchange". See Barkow *et al.* (1992), 163–228.

Cosmides, L. & J. Tooby 1997. "Evolutionary Psychology: A Primer". www.psych.ucsb.edu/research/cep/primer.html (accessed May 2007).

Cummins, R. [1975] 1994. "Functional Analysis". See Sober (1994), 49–70. Originally published in *Journal of Philosophy* **72** (1975), 741–64.

Darwin, C. [1859] 1968. *The Origin of Species*. Reprint of the 1st edn, W. J. Burrow (ed.). Harmondsworth: Penguin.

Darwin, C. [1871] 1981. *The Descent of Man, and Selection in Relation to Sex*. Reprint of the First Edition. Princeton, NJ: University of Princeton Press.

Darwin, F. & A. C. Seward (ed.) 1903. *More Letters of Charles Darwin*, 2 vols. London: John Murray.

Davidson, D. [1974] 1984. "On the Very Idea of a Conceptual Scheme". Reprinted in his *Inquiries into Truth and Interpretation*. Oxford: Clarendon Press.

Dawkins, R. 1979. "In Defence of Selfish Genes". *Philosophy* **56**: 556–73.

Dawkins, R. 1982. *The Extended Phenotype*. Oxford: Oxford University Press.

Dawkins, R. 1983. "Universal Darwinism". In *Evolution from Molecules to Man*, D. S. Bendall (ed.), 403–25. Cambridge: Cambridge University Press. Reprinted in Hull & Ruse (1998), 15–37.

Dawkins, R. 1986. *The Blind Watchmaker*. Harmondsworth: Penguin.

Dawkins, R. [1976] 1989. *The Selfish Gene*, 2nd edn. Oxford: Oxford University Press.

Dawkins, R. 1993. "Viruses of the Mind". In *Dennett and his Critics*, B. Dahlbom (ed.), 13–27. Oxford: Blackwell.

Dawkins, R. 2004. *The Ancestor's Tale: A Pilgrimage to the Dawn of Life*. London: Weidenfeld & Nicolson.

Dennett, D. 1987. "Intentional Systems in Cognitive Ethology: The 'Panglossian Paradigm' Defended". In his *The Intentional Stance*. Cambridge, MA: MIT Press.

Dennett, D. 1995. *Darwin's Dangerous Idea: Evolution and the Meanings of Life*. Harmondsworth: Penguin.

Dennett, D. 2006. *Breaking the Spell: Religion as a Natural Phenomenon*. Harmondsworth: Allen Lane.

Dupré, J. (ed.) 1987. *The Latest on the Best: Essays on Evolution and Optimality*. Cambridge, MA: MIT Press.

Dupré, J. 1993. *The Disorder of Things*. Cambridge, MA: Harvard University Press.

Edelman, G. 1987. *Neural Darwinism: The Theory of Neuronal Group Selection*. New York: Basic Books.

Eldredge, N. & S. J. Gould 1972. "Punctuated Equilibria". In *Models in Paleobiology*, T. M. Schopf (ed.), 82–115. San Francisco, CA: Freeman, Cooper.

Ereshefsky, M. (ed.) 1992. *The Units of Evolution: Essays on the Nature of Species*. Cambridge, MA: MIT Press.

Ereshefsky, M. 2000. *The Poverty of the Linnaean Hierarchy: A Philosophical Study of Biological Taxonomy*. Cambridge: Cambridge University Press.

Feyerabend, P. 1993. *Against Method: Outline of an Anarchistic Theory of Knowledge*, 3rd edn. London: Verso.

Fodor, J. 2000. *The Mind Doesn't Work That Way*. Cambridge, MA: MIT Press.

Freud, S. [1917] 1963. *Introductory Lectures on Psychoanalysis*, part III. London: Hogarth Press.

Futuyma, D. J. 1998. *Evolutionary Biology*, 3rd edn. Sunderland, MA: Sinauer.

Ghiselin, M. 1974. "A Radical Solution to the Species Problem". *Systematic Zoology* **23**: 536–44.

Gilbert, S. 2003. *Developmental Biology*, 7th edn. Sunderland, MA: Sinauer.

Glannon, W. 1997. "Psychopathy and Responsibility". *Journal of Applied Philosophy* **14**(3): 263–75.

Glendinning, S. 1996. "Heidegger and the Question of Animality". *International Journal of Philosophical Studies* **4**(1): 67–86.

Gould, S. J. 1977. *Ontogeny and Phylogeny*. Cambridge, MA: Belknap.

Gould, S. J. 1978. *Ever Since Darwin: Reflections in Natural History*. Harmondsworth: Penguin.

Gould, S. J. 1980. "The Panda's Thumb". In his *The Panda's Thumb: More Reflections in Natural History*, 19–27. Harmondsworth: Penguin.

Gould, S. J. 1991. "Exaptation: A Crucial Tool for Evolutionary Psychology". *Journal of Social Issues* **47**: 43–65.

Gould, S. J. 2000. *Wonderful Life*. London: Vintage.

Gould, S. J. 2002. *The Structure of Evolutionary Theory*. Cambridge, MA: Harvard University Press.

Gould, S. J. & R. Lewontin 1979. "The Spandrels of San Marco and the Panglossian Paradigm: A Critique of the Adaptationist Programme". *Proceedings of the Royal Society of London* B205: 581–98. Reprinted in Sober (1994), 73–90; (2006), 79–98.

Gould, S. J. & E. Vrba 1982. "Exaptation – A Missing Term in the Science of Form". *Paleobiology* **8**(1): 4–15. Reprinted in Hull & Ruse (1998), 52–71.

Grant, V. 1981. *Plant Speciation*, 2nd edn. New York: Columbia University Press.

Grantham, T. & S. Nichols 1999. "Evolutionary Psychology: Ultimate Explanations and Panglossian Predictions". See Hardcastle (1999), 47–66.

Gray, J. 2003. *Straw Dogs: Reflections on Humans and Other Animals*. London: Granta.

Griffiths, P. 2002. "What is Innateness?" *The Monist* **85**(1): 70–85.

Griffiths, P. 2007. "Evo-Devo Meets the Mind: Towards a Developmental Evolutionary Psychology". In *Integrating Evolution and Development*, R. Brandon & R. Sanson (eds), Cambridge, MA: MIT Press.

Griffiths, P. & R. Gray 1994. "Developmental Systems and Evolutionary Explanation". *Journal of Philosophy* 91(6): 277–304. Reprinted in Hull & Ruse (1998), 117–46.

Hahlweg, K. & C. A. Hooker (ed.) 1989. *Issues in Evolutionary Epistemology*. Albany, NY: SUNY Press.

Hamilton, W. 1964a. "The Genetical Evolution of Social Behaviour I". *Journal of Theoretical Biology* 7: 1–16.

Hamilton, W. 1964b. "The Genetical Evolution of Social Behaviour II". *Journal of Theoretical Biology* 7: 17–52.

Hardcastle, V. G. (ed.) 1999. *Where Biology Meets Psychology*. Cambridge, MA: MIT Press.

Hennig, W. 1966. *Phlyogenetic Systematics*. Urbana, IL: University of Illinois Press.

Hobsbawm, E. 1994. *Age of Extremes: The Short Twentieth Century 1914–1991*. London: Abacus.

Hodge, J. & G. Radick (ed.) 2003. *The Cambridge Companion to Darwin*. Cambridge: Cambridge University Press.

Hofstadter, D. 1985. *Metamagical Themas: Questing for the Essence of Mind and Pattern*. New York: Basic Books.

Hull, D. L. [1978] 1994. "A Matter of Individuality". See Sober (1994), 193–216. Originally published in *Philosophy of Science* 45 (1978): 335–60.

Hull, D. L. & M. Ruse (eds) 1998. *The Philosophy of Biology*. Oxford: Oxford University Press.

Hume, D. [1748] 1975. *An Enquiry Concerning Human Understanding*, L. A. Selby-Bigge (ed.), P. H. Nidditch (rev.). Oxford: Oxford University Press.

Hume, D. [1739–40] 1978. *A Treatise of Human Nature*, L. A. Selby-Bigge (ed.), 2nd edn revised by P. H. Nidditch. Oxford: Oxford University Press.

Hume, D. [1779] 1998. *Dialogues Concerning Natural Religion, and the Natural History of Religion*, J. C. A. Gaskin (ed.). Oxford: Oxford University Press.

James, W. [1890] 1981. *The Principles of Psychology* [3 vols]. In *The Works of William James*. Cambridge, MA: Harvard University Press.

Jenkin, F. 1867. "The Origin of Species". *North British Review* 46: 277–318.

Kaplan, S. 1992. "Environmental Preference in a Knowledge-Seeking, Knowledge-Using Organism". See Barkow *et al.* (1992), 581–98.

Kaufmann, S. 1995. *At Home in the Universe: The Search for the Laws of Self-Organization and Complexity*. Oxford: Oxford University Press.

Keeley, B. 2004. "Anthropomorphism, Primatomorphism, Mammalomorphism: Under-standing Cross-Species Comparisons". *Biology and Philosophy* 19(4): 521–40.

Keller, L. & K. G. Ross 1998. "Selfish Genes: A Green Beard in the Red Fire Ant". *Nature* 394: 573–5.

Kimura, M. 1982. *The Neutral Theory of Molecular Evolution*. Cambridge: Cambridge University Press.

Kripke, S. 1980. *Naming and Necessity*, rev. edn. Oxford: Blackwell.

Lazzaro, B. P., T. B. Sackton, A. G. Clark 2006. "Genetic Variation in *Drosophila melanogaster* Resistance to Infection: A Comparison Across Bacteria". *Genetics* 174: 1539–54.

LeCroy, D. 2000. "Freud: The First Evolutionary Psychologist?" *Annals of the New York Academy of Sciences* 907: 182–90.

Lehrman, D. 1953. "A Critique of Konrad Lorenz's Theory of Instinctive Behaviour". *Quarterly Review of Biology* 28: 337–63.

Leslie, J. 1989. *Universes*. London: Routledge.

Lewens, T. 2005. *Organisms and Artifacts: Design in Nature and Elsewhere*. Cambridge, MA: Bradford Books.

Lewens, T. 2006. *Darwin*. London: Routledge.

Lewontin, R. 1972. "The Apportionment of Human Diversity". *Evolutionary Biology* 6: 381–98.

Lewontin, R. 1983. "Gene, Organism and Environment". In *Evolution from Molecules to Man*, D. S. Bendall (ed.), 273–85. Cambridge: Cambridge University Press.

Lewontin, R. 1993. *The Doctrine of DNA: Biology as Ideology*. Harmondsworth: Penguin.

Lewontin, R. 2000a. *The Triple Helix: Gene, Organism and Environment*. Cambridge, MA: Harvard University Press.

Lewontin, R. 2000b. "The Inferiority Complex". In his *It Ain't Necessarily So*. London: Granta.

Locke, J. [1690] 1979. *An Essay Concerning Human Understanding*, P. H. Nidditch (ed.). Oxford: Clarendon Press.

Lorenz, K. 1950. "The Comparative Method in Studying Innate Behavior Patterns". *Symposia of the Society of Experimental Biology* 4: 221–68.

Lorenz, K. 1965. *Evolution and Modification of Behavior*. Chicago, IL: University of Chicago Press.

Lucretius 1916. *On the Nature of Things*, W. E. Leonard (trans.). London: J. M. Dent.

Mackie, J. L. 1978 "The Law of the Jungle". *Philosophy* 53: 455–64.

Mackie, J. L. 1981. "Genes and Egoism". *Philosophy* 56(218): 553–5.

Maclaurin, J. 2002. "The Resurrection of Innateness". *The Monist* 85(1): 105–30.

Manning, G. 2006. *A Quick and Simple Introduction to* Drosophila melanogaster. www.ceolas. org/fly/intro.html (accessed July 2007).

Manson, N. 2003. *God and Design: The Teleological Argument and Modern Science*. London: Routledge.

Martin, R. D. 1981. "Relative Brain Size and Basal Metabolic Rate in Terrestrial Vertebrates". *Nature* 293: 57–60.

Maynard Smith, J. 1993. *The Theory of Evolution*. Cambridge: Canto.

Mayr, E. 1942. *Systematics and the Origin of Species from the Viewpoint of a Zoologist*. New York: Columbia University Press (reprinted Cambridge, MA: Harvard University Press, 1999).

Mayr, E. 1988. *Toward a New Philosophy of Biology: Observations of an Evolutionist*. Cambridge, MA: Belknap.

Mayr, E. 1993. *One Long Argument: Charles Darwin & the Genesis of Modern Evolutionary Thought*. Cambridge, MA: Harvard University Press.

McFarland, D. (ed.) 1987. *The Oxford Companion to Animal Behaviour*. Oxford: Oxford University Press.

McGinn, C. 1991. *The Problem of Consciousness*. Oxford: Blackwell.

McGinn, C. 1999. *The Mysterious Flame: Conscious Minds in a Material World*. New York: Basic Books.

Melville, H. [1851] 1972. *Moby-Dick; or, The Whale*, H. Beaver (ed.). Harmondsworth: Penguin.

Midgley, M. 1979. "Gene-Juggling". *Philosophy* 54: 439–58.

Miller, K. 1999. *Finding Darwin's God: A Scientist's Search for Common Ground Between God and Evolution*. New York: Cliff Street.

Miller, K. 2003. "Answering the Biochemical Argument from Design. In *God and Design: The Teleological Argument and Modern Science*, N. Manson (ed.), 292–307. London: Routledge.

Miller, K. 2004. "The Flagellum Unspun: The Collapse of Irreducible Complexity". In *Debating Design: From Darwin to DNA*, M. Ruse & W. Dembski (ed.), 81–97. Cambridge: Cambridge University Press.

Millikan, R. 1984. *Language, Thought and Other Biological Categories*. Cambridge, MA: MIT Press.

Millikan, R. 1989. "In Defense of Proper Functions". *Philosophy of Science* 56: 288–302.

Mishler, B. & R. Brandon [1987] 1998. "Individuality, Pluralism and the Phylogenetic Species Concept". See Hull & Ruse (1998), 300–318. Originally published in *Biology and Philosophy* 2 (1978): 397–414.

Mishler, B. & M. Donoghue [1982] 1994. "Species Concepts: A Case for Pluralism". See Sober (1994), 217–32. Originally published in *Systematic Zoology* 31 (1982): 491–503.

Moore, G. E. [1903] 1993. *Principia Ethica*, T. Baldwin (ed.). Cambridge: Cambridge University Press.

Morris, D. 1967. *The Naked Ape*. London: Jonathan Cape.

Okasha, S. 2001. "Why Won't the Group Selection Controversy Go Away?" *British Journal for the Philosophy of Science* 52: 25–50.

Orians, G. H. & J. H. Heerwagen 1992. "Evolved Responses to Landscapes". See Barkow *et al.* (eds) (1992), 555–80.

Orzack, S. H. & E. Sober (eds) 2001. *Adaptationism and Optimality*. Cambridge: Cambridge University Press.

Oyama, S. 1985. *The Ontogeny of Information*. Cambridge: Cambridge University Press.

Oyama, S. 2000. *Evolution's Eye: A Systems View of the Biology–Culture Divide*. Durham, NC: Duke University Press.

Oyama, S., P. Griffiths, R. Gray (eds) 2001. *Cycles of Contingency: Developmental Systems and Evolution*. Cambridge, MA: MIT Press.

Paley, W. 1809. *Natural Theology: or, Evidences of the Existence and Attributes of the Deity*, 12th edn. London: J. Faulder.

Parr, L. A. & F. B. M. de Waal 1999. "Visual Kin Recognition in Chimpanzees". *Nature* 399 (17 June 1999): 647–8.

Pinel, J. J. 2000. *Biopsychology*, 4th edn. Boston, MA: Allyn & Bacon.

Pinker, S. 1997. *How the Mind Works*. Harmondsworth: Penguin.

Plantinga, A. 1991. "When Faith and Reason Clash: Evolution and the Bible. *Christian Scholar's Review* 21: 8–32.

Popper, K. 1974. "Replies to my Critics". In *The Philosophy of Karl Popper*, vol. 2, P. A. Schlipp (ed.), 961–1180. LaSalle, IL: Open Court.

Popper, K. 1976. *Unended Quest: An Intellectual Autobiography*. London: Fontana.

Popper, K. 1979. *Objective Knowledge: An Evolutionary Approach*, 2nd edn. Oxford: Oxford University Press.

Putnam, H. 1975. "What is Mathematical Truth?" In his *Mathematics, Matter and Method: Philosophical Papers vol. 1*. Cambridge: Cambridge University Press.

Queller, D. C. 1995. "The Spaniels of St Marx and the Panglossian Paradox: A Critique of a Rhetorical Programme". *Quarterly Review of Biology* 70(4): 485–9.

Quicke, D. L. J. 1993. *Principles and Techniques of Contemporary Taxonomy*. Glasgow: Blackie.

Quine, W. V. 1961. "Two Dogmas of Empiricism". In his *From a Logical Point of View*, 2nd edn, 20–46. Cambridge, MA: Harvard University Press.

Quine, W. V. 1969. *Ontological Relativity and Other Essays*. New York: Columbia University Press.

Quine, W. V. 1995. *From Stimulus to Science*. Cambridge, MA: Harvard University Press.

Radick, G. 2000. "Two Explanations of Evolutionary Progress". *Biology and Philosophy* 15: 475–91.

Rees, M. 1999. *Just Six Numbers: The Deep Forces that Shape the Universe*. London: Weidenfeld & Nicolson.

Richards, J. R. 2000. *Human Nature After Darwin*. London: Routledge.

Ridley, M. 1989. *Evolution and Classification: The Reformation of Cladism*. New York: John Wiley.

Ridley, M. (ed.) 1997. *Evolution*. Oxford: Oxford University Press.

Ridley, M. 1996. *The Origins of Virtue*. Harmondsworth: Penguin.

Ridley, M. 2003a. *Evolution*, 3rd edn. Oxford: Blackwell.

Ridley, M. 2003b. *Nature via Nurture: Genes, Experience and what makes us Human*. London: HarperCollins.

Rose, H. & S. Rose (eds) 2000. *Alas, Poor Darwin: Arguments Against Evolutionary Psychology*. London: Jonathan Cape.

Rose, S. 1978. "Pre-Copernican Sociobiology?" *New Scientist* **80**: 45–6.

Rosen, D. E. 1978. "Vicariant Patterns and Historical Explanations in Biogeography". *Systematic Zoology* 27: 159–88.

Rosenberg, A. 1994. *Instrumental Biology or the Disunity of Science*. Chicago, IL: University of Chicago Press.

Rottschaefer, W. & D. Martinsen [1990] 1995. "Really Taking Darwin Seriously: An Alternative to Michael Ruse's Darwinian Metaethics". See Thompson (1995), 375–408.

Ruse, M. [1986] 1995. "Evolutionary Ethics: A Phoenix Arisen". See Thompson (1995), 225–48.

Ruse, M. 1999. *The Darwinian Revolution: Science Red in Tooth and Claw*, 2nd edn. Chicago, IL: University of Chicago Press.

Ruse, M. 2004. *Can a Darwinian Be a Christian?*, 2nd edn. Cambridge: Cambridge University Press.

Sampson, G. 1997. *Educating Eve: The "Language Instinct" Debate*. London: Cassell.

Schank, J. C. & W. C. Wimsatt 1986. "Generative Entrenchment and Evolution". *Proceedings of the Biennial Meeting of the Philosophy of Science Association* 1986, 2: Symposia and Invited Papers, 33–60.

Schwartz, J. H. 1999. *Sudden Origins: Fossils, Genes, and the Emergence of Species*. New York: John Wiley.

Segerstråle, U. 2000. *Defenders of the Truth: The Sociobiology Debate*. Oxford: Oxford University Press.

Sesardic, N. 1995. "Recent Work on Human Altruism and Evolution". *Ethics* **106**: 128–57.

Sheldon Davies, P. 1999. "The Conflict of Evolutionary Psychology". See Hardcastle (1999): 67–82.

Sheppard, R. N. 1992. "The Perceptual Organization of Colors: An Adaptation to Regularities in the Terrestrial World?" See Barkow *et al.* (eds) (1992), 495–532.

Simpson, G. G. 1961, *Principles of Animal Taxonomy*. New York: Columbia University Press.

Skelton, P. (ed.) 1993. *Evolution: A Biological and Palaeontological Approach*. Wokingham: Addison-Wesley.

Skyrms, B. 1996. *Evolution of the Social Contract*. Cambridge: Cambridge University Press.

Sneath, P. H. A. & R. R. Sokal 1973. *Numerical Taxonomy*. San Francisco, CA: W. H. Freeman.

Sober, E. 1984. *The Nature of Selection*. Cambridge, MA: MIT Press.

Sober, E. (ed.) 1994. *Conceptual Issues in Evolutionary Biology*, 2nd edn. Cambridge, MA: MIT Press.

Sober, E. (ed.) 2006. *Conceptual Issues in Evolutionary Biology*, 3rd edn. Cambridge, MA: MIT Press.

Sober, E. & D. S. Wilson 1998. *Unto Others*. Cambridge, MA: Harvard University Press.

Sperber D., F. Cara, V. Girotto 1995. "Relevance Theory Explains the Selection Task". *Cognition* 57(1): 31–95.

Stebbins, G. L. 1982. *Darwin to DNA, Molecules to Humanity*. San Francisco, CA: W. H. Freeman.

Sterelny, K. 2001. *Dawkins vs. Gould: Survival of the Fittest*. Cambridge: Icon.

Sterelny, K. & P. Kitcher 1988. "The Return of the Gene". *Journal of Philosophy* 85(7): 339–61.

Sterelny, K., K. C. Smith, M. Dickison 1996. "The Extended Replicator". *Biology and Philosophy* 11(3): 377–403.

Stern, C. (ed.) 2004. *Gastrulation: From Cells to Embryo*. Cold Spring Harbor, NY: Cold Spring Harbor Laboratory Press.

Stich, S. 1975. *Innate Ideas*. Berkeley, CA: University of California Press.

Swinburne, R. 1991. *The Existence of God*, rev. edn. Oxford: Clarendon Press.

Symons, D. 1992. "On the Use and Misuse of Darwinism in the Study of Human Behavior". See Barkow *et al.* (eds) (1992), 137–59.

Tao, Y., D. L. Hartl, C. C. Laurie 2001. "Sex-ratio Segregation Distortion Associated with Reproductive Isolation in Drosophila". *Proceedings of the National Academy of Sciences of the United States of America* 98: 13183–8.

Thompson, P. (ed.) 1995. *Issues in Evolutionary Ethics*. New York: SUNY Press.

Thomson, W. (Lord Kelvin) 1862. "On the Age of the Sun's Heat". *Macmillan's Magazine* 5 (5 March): 288–93. Reprinted in Kelvin's *Popular Lectures and Addresses*, vol. 1 (London: Macmillan 1891).

Thornhill, R. & C. R. Palmer 2000. *A Natural History of Rape*. Cambridge, MA: MIT Press.

Tooby, J. & L. Cosmides 1992. "The Psychological Foundations of Culture". See Barkow *et al.* (eds) (1992), 19–136.

Tudge, C. 2000. *The Variety of Life: A Survey and Celebration of All the Creatures that have Ever Lived*. Oxford: Oxford University Press.

Vokey, J. R., D. Rendell, J. M. Tangen *et al.* 2004. "Visual Kin Recognition and Family Resemblance in Chimpanzees". *Journal of Comparative Psychology* 118(2): 194–9.

Voltaire, François Marie Arouet [1759] 1937. *Candide and Other Tales*, T. Smollett (trans.), J. C. Thornton (rev.). London: Dent.

Waddington, C. H. 1962. *New Patterns in Genetics and Development*. New York: Columbia Unviversity Press.

Waddington, C. H. 1975. *The Evolution of an Evolutionist*. Edinburgh: Edinburgh University Press.

Williams, G. 1966. *Adaptation and Natural Selection*. Princeton, NJ: Princeton University Press.

Williams, G. C. & R. Nesse 1991. "The Dawn of Darwinian Medicine". *Quarterly Review of Biology* 66(1): 1–22.

Wilson, D. S., E. Dietrich, A. B. Clark 2003. "On the Inappropriate Use of the Naturalistic Fallacy in Evolutionary Psychology". *Biology and Philosophy* 18(5): 669–82.

Wilson, E. O. 1975. *Sociobiology*. Cambridge, MA: Belknap.

Wilson, E. O. 1978. *On Human Nature*. Cambridge, MA: Harvard University Press.

Wilson, E. O. 1995. "Science and Ideology". *Academic Questions* 8(3): 73–81.

Wilson, M. & M. Daly 1992. "The Man who Mistook his Wife for a Chattel". See Barkow *et al.* (eds) (1992), 289–322.

Wilson, R. A. (ed.) 1999. *Species: New Interdisciplinary Essays*. Cambridge, MA: MIT Press.

Wimsatt, W. 1980. "The Units of Selection and the Structure of the Multi-Level Genome". *Proceedings of the Biennial Meeting of the Philosophy of Science Association, 1980*, vol. 2 [Symposia and Invited Papers]: 122–83.

Wimsatt, W. 1999. "Generativity, Entrenchment, Evolution, Innateness". See Hardcastle (1999), 139–80.

Woodward, J. 2002. "There is No Such Thing as a *Ceteris Paribus* Law". *Erkenntnis* 57(3): 303–28.

Wright, L. [1973] 1994. "Functions". See Sober (1994), 27–48. Originally published in *Philosophical Review* 82 (1973): 139–68.

Index